Fundamentals of Fluidized-bed Chemical Processes

Butterworths Monographs in Chemical Engineering

Butterworths Monographs in Chemical Engineering is a series of occasional texts by internationally acknowledged specialists, providing authoritative treatment of topics of current significance in chemical engineering.

Series Editor

J W Mullin
Professor of Chemical Engineering, University College, London

Published titles:
Solid-liquid separation
Liquids and liquid mixtures—3rd edn

Forthcoming titles:
Enlargement and compaction of particulate solids
Mixing in the process industries
Diffusion and heat flow in liquids
Introduction to electrode materials

Butterworths Monographs in Chemical Engineering

Fundamentals of Fluidized-bed Chemical Processes

J G Yates, PhD, DIC, CChem, FRSC
Department of Chemical and Biochemical Engineering,
University College, London

Butterworths
London Boston Durban Singapore Sydney Toronto Wellington

First published 1983

British Library Cataloguing in Publication Data

Yates, J.G.
Fundamentals of fluidized-bed chemical processes.
—(Butterworths monographs in chemistry)
1. Fluidization
I. Title
660.2'842 TP156.F65

ISBN 0-408-70909-X

Typeset by Scribe Design, Gillingham, Kent
Printed by The Thetford Press Ltd, Thetford, Norfolk

Preface

Fluidized-bed reactors have been used on an industrial scale for over fifty years and during this time an enormous amount of work has been carried out on all aspects of their design and operation. This work has generated a correspondingly large body of literature which is being augmented at a steady rate. An author of a review of fluidized-bed chemical processes is therefore faced with the dilemma of either overloading his readers with a comprehensive coverage of the world literature on the subject or providing a more manageable survey which runs the risk of being superficial. I have tried in this book to steer a course somewhere between these two obstacles and to present the fundamentals of the subject in a suitably digestible form, but I acknowledge that in so doing I have been obliged to omit reference to much of the work that has been done in the field of fluidization in many industrial and university laboratories around the world. My hope is that this 'biased model' will not prove too unacceptable to those whose work I have treated too lightly or not at all.

Following a short introduction the book is divided into five main chapters. The first deals with the basic physics of the subject while the second shows how the physics may be combined with chemical kinetics to generate models of fluidized-bed reactors. The next two chapters deal, again in a fundamental way, with two major applications of the technique, one the well-established Fluidized Catalytic Cracking process, the other a newer development concerned with the combustion and gasification of coal and heavy oils. The last chapter examines a number of miscellaneous processes that have been and are currently used for the production of chemicals such as phthalic anhydride, acrylonitrile and compounds of uranium.

Although the prime aim of the book is to present a concise, up-to-date picture of the chemically-orientated aspects of gas-solid fluidization suitable for final-year undergraduate and postgraduate students of chemical engineering, it is hoped that practising engineers and scientists involved with gas-solid systems will also find something of value within its pages.

It would not be fitting to end this Preface without expressing gratitude to all those in the fluidization field whom I have come to know and respect during the last fifteen years and on whose work and ideas I have so freely drawn. If I were to select one name for special thanks from those liberally distributed throughout the text it would be that of Peter Rowe who has led the fluidization research group at UCL since 1965 and who has never failed to stimulate and inspire all those who have worked with him. A final word of thanks goes to John Mullin at whose suggestion I wrote this book and without whose encouragement and help it would not have seen the light of day.

J.G. Yates

University College
London

The poet, the painter, the scientist, each superimposes his more or less ephemeral vision on the universe, each constructs his own biased model of reality by selecting and highlighting those aspects of experience which he considers significant and ignoring those which he considers irrelevant.

Arthur Koestler

Contents

Nomenclature

The list of symbols given below is not exhaustive, some having been omitted because they are only used once or twice in specialist sections of the book. All symbols are however defined in the text and so no confusion should arise from the omissions. SI units have been used wherever possible although in many cases temperatures are given in degrees Celsius and pressures in bar.

A	cross sectional area of bed	m^2
A_c	area of bed occupied by bubble–cloud phase	m^2
A_e	area of bed occupied by emulsion phase	m^2
a	catalyst activity	–
C_A	concentration of reactant A	mol/m^3
C_{Ao}	initial concentration of reactant A	mol/m^3
C_D	drag coefficient	–
C_{ps}	specific heat capacity of solid	kJ/kg K
$\dot{C}$	carbon burning rate	mol/s
D	diameter of bed	m
D_G	molecular diffusion coefficient	m^2/s
d_a	diameter of sieve aperture	m
d_b	bubble diameter	m
d_c	cloud diameter	m
d_i	initial diameter of particle	m
d_p	particle diameter	m
E	diffusion coefficient	m^2/s
F	mass flux of solids	kg/m^2s
F_o	solids flowrate at bed surface	kg/m^2s
F_∞	solids elutriation rate	kg/m^2s
f_b	fraction of bed occupied by bubbles	–
f_c	mass fraction of feed deposited on catalyst as coke	–
f_w	fraction of bubble sphere occupied by wake	–
f_s	fraction of bed occupied by slugs	–
Ga	Galileo number, $\dfrac{d_p^3\rho_g(\rho_s - \rho_g)g}{\mu^2}$	–
Ha	Hatta number, $\dfrac{(kD_G)^{1/2}}{k_g}$	–

H_{mf}	bed height at minimum fluidization	m
h	bed height above distributor	m
h_{bs}	bed-to-surface heat transfer coefficient	W/m^2 K
K_{bc}	gas exchange rate between bubble and cloud per unit bubble volume	s^{-1}
K_{bce}	gas exchange rate between bubble–cloud and emulsion per unit bubble volume	s^{-1}
K_{be}	gas exchange rate between bubble and emulsion per unit bubble volume	s^{-1}
K_{ce}	gas exchange rate between cloud and emulsion per unit bubble volume	s^{-1}
K^*	elutriation rate coefficient	kg/m^2 s
k	reaction rate coefficient	s^{-1}
k_{be}	bubble–emulsion mass transfer coefficient	m/s
k_d	rate coefficient of catalyst deactivation	s^{-1}
k_{gc}	cloud–emulsion mass transfer coefficient	m/s
k_g	mass transfer coefficient	m/s
k_s	surface reaction rate coefficient	m/s
k_v	velocity coefficient	–
L	length	m
M	mixing index	–
N_α	number of transfer units	–
N_E	number of mixing units	–
N_r	number of reaction units	–
n_s	slug frequency	s^{-1}
p	pressure	kN/m^2
Q	volumetric gas flowrate	m^3/s
Q_b	bubble flowrate	m^3/s
Q_{mf}	gas flowrate at minimum fluidization	m^3/s
Q_s	slug flowrate	m^3/s
Q_T	total volumetric gas flowrate	m^3/s
R_b	bubble radius	m
R_c	cloud radius	m
Re	Reynolds number, $\dfrac{U d_p \rho_g}{\mu}$	–
Re_{mf}	Reynolds number at minimum fluidization, $\dfrac{U_{mf} d_p \rho_g}{\mu}$	–
Re_t	Reynolds number under terminal fall conditions, $\dfrac{U_t d_p \rho_g}{\mu}$	–
r	radial distance	m
S	surface area	m^2
Sc	Schmidt number, $\dfrac{\mu}{\rho_g D_G}$	–

Sh	Sherwood number, $\frac{k_g d_p}{D_G}$	–
T	temperature	K
t	time	s
$\bar{t}$	mean residence time	s
t_c	burn-out time of batch	s
U	velocity	m/s
U_b	bubble belocity	m/s
U_c	cloud velocity	m/s
U_e	emulsion phase gas velocity	m/s
U_f	fluid velocity	m/s
U_{mb}	minimum bubbling velocity	m/s
U_{mf}	superficial velocity at minimum fluidization	m/s
U_p	particle velocity	m/s
U_s	minimum slugging velocity	m/s
U_{sl}	slip velocity	m/s
U_t	particle terminal fall velocity	m/s
U_{TO}	take-over velocity	m/s
V	volume	m^3
V_b	bubble volume	m^3
V_{bc}	volume of gas in bubble plus cloud	m^3
V_c	cloud volume	m^3
X	interphase mass transfer term	–
x	mass fraction	–
z	vertical coordinate	m
α	ratio of bubble velocity to interstitial gas velocity	–
β	calcium-to-sulphur mole ratio	–
δ	film thickness	m
ε	voidage	–
ε_{mf}	voidage at minimum fluidization	–
θ	angle	degrees
μ	fluid viscosity	Ns/m^2
υ	kinematic viscosity	m^2/s
ρ	density	kg/m^3
ρ_f	fluid density	kg/m^3
ρ_g	gas density	kg/m^3
ρ_s	solids density	kg/m^3
ϕ	sphericity factor	–
ψ_f	fluid stream function	–

Introduction

The technique of gas-solid fluidization was first used industrially in the Winkler process for the gasification of coal in the early 1930's, but for various reasons the process did not find widespread use and the technique was not developed further until the beginning of the Second World War. Then a group of companies, which included Standard Oil New Jersey, M.W. Kellogg, Shell and Universal Oil Products, in an effort designed primarily to exploit the catalytic cracking method for gasoline manufacture discovered by the French engineer Eugene Houdry but without using his fixed-bed reactors, designed a fluidized solids process that has formed the foundation of all subsequent developments in the field[1].

The Fluidized Catalytic Cracking (FCC) process was a spectacularly successful example of engineering innovation and it remains today one of the cornerstones of petroleum refining technology. Following the success of the FCC process a number of other processes using fluidized solids were put into commercial operation and although some of these were successful others were not and never operated to their full design specifications. It became clear that there were certain features of the FCC and similar processes that made them particularly suited to application in fluidized beds. That other processes were not so well suited was found to be a result of their having one or more of a number of unfavourable characteristics associated with, for example, the chemistry of the reactions, the properties of the solids or the pattern of gas-solid contact, and as a result of certain plant failures fluidized beds acquired a reputation for unreliability.

The fluidized bed is only one of the many types of reactor employed in industry for carrying out gas-solid reactions but it has a number of advantages over its competitors that are worth noting at the outset. Its most important advantages stem from the fact that the solid particles it contains are in continuous motion and are normally very well mixed; the result is that 'hot spots' are rapidly dissipated and the bed operates in an essentially isothermal manner.

Furthermore, because of the very high bed-to-surface heat transfer that can be achieved, again as a result of particle motion, temperature control is seldom a problem. Thirdly, the fluid-like properties of the gas-solid mixture enable the solid to be transferred without difficulty from one vessel to another, a useful feature in cases where the solid is required to take part in two different reactions as part of the same overall process. There are however a number of disadvantages of fluidized beds that can be so severe in some cases as to prevent their use. Amongst these are (i) the erosion of bed internals, heat transfer coils, valves etc. caused by the 'sand blasting' action of the solids, (ii) the loss of very fine particles through cyclone plugging etc. (a vital consideration if the solid is an expensive catalyst), (iii) the by-passing of the solid by gas bubbles which can severely limit the conversion on a once through basis. Another problem is that owing to our limited understanding of the physics of fluidization reliable scale-up from laboratory reactor to full-scale plant can be difficult to achieve.

Despite these problems fluidized-bed reactors are widely used by industry. Zenz[2] listed over 100 examples of fluid-bed applications up to 1978 and commented that "fluidization as a unit operation has touched almost every process industry". A few examples will serve to illustrate the breadth of these applications and to show some of the reasons for choosing a fluidized bed rather than another more conventional alternative.

The FCC process already mentioned and to be discussed in detail later (Chapter 3) is successful essentially because it enables high throughputs to be maintained on a continuous basis and overcomes, in an elegant and economic way, the severe problem of catalyst deactivation. Control of reaction temperature and good heat transfer are factors which recommend fluidized beds for highly exothermic processes such as the chlorination of hydrocarbons and the ammoxidation of propylene to acrylonitrile. The latter involves the reaction:

$$CH_3CH{=}CH_2 + NH_3 + 1\tfrac{1}{2}O_2 = CH_2{=}CHCN + 3H_2O;$$
$$\Delta H = -515 \text{kJ/mol}$$

as well as a number of others leading to by-products which can give a total heat release of some 700 kJ/mol of propylene reacted. Nevertheless in the Sohio fluidized-bed process, by means of which almost the whole world supply of acrylonitrile is made, the reactors are readily controlled at 400–500°C for long runs and with good heat recovery.

The good solids mixing characteristics of fluidized beds have proved of overriding importance in, for example, the Dorr-Oliver FluoSolids process where iron pyrites is roasted to give the metal oxide and SO_2:

$$4FeS_2 + 11O_2 = 2Fe_2O_3 + 8SO_2$$

and in the production of titanium tetrachloride from ilmenite where an intimate mixture of the ore and a carbon-containing reducing agent are fluidized with chlorine;

$$TiO_2FeO + 3\tfrac{1}{2}Cl_2 + 2C = TiCl_4 + FeCl_3 + CO + CO_2$$

An ingeneous process that made use of a number of the foregoing advantages of the fluidized solids technique was that operated for a number of years by the Arad Chemical Company at Mishor Rotem, Israel[3]. The process was designed to produce large tonnages of hydrochloric acid to be used at a later stage in combination with phosphate rock for the production of fluoride-free phosphoric acid. The hydrochloric acid was made by the hydrolysis of magnesium chloride derived from the "end brines" of the Dead Sea salt-extraction process:

$$MgCl_2 + H_2O = MgO + 2HCl$$

the reaction being carried out in large refractory-lined beds of sand fluidized with air. The system was maintained at a temperature in the region of 900°C by burning crude oil in the bed, and in the course of the reaction the solid magnesium oxide product was deposited on the sand particles as a durable silicate coating while the gaseous HCl was removed overhead via cyclones and subjected to further treatment. Both oil and brine were injected directly into the bed where, following rapid evaporation, the oil burned and the brine solids were dispersed due to the churning motion of the bed particles. Despite its many attractive features however the process was plagued by technical and economic problems which eventually became so severe as to cause the plant, at enormous cost, to be shut down.

Failures of this kind have usually been attributed to a lack on the part of the plant designers of a sufficient understanding of the basic physics of gas-solid fluidization and during the last three decades a great deal of fundamental research has been carried out with the aim of clarifying these underlying features. It is with the results of much of this research that we will be concerned in this book.

References

1 JAHNIG C.E., CAMPBELL D.L. AND MARTIN H.Z., *Fluidization*, Grace, J.R., and MATSEN J.M. (eds), 3 Plenum Press, New York (1980
2 ZENZ F.A., *Encyclopaedia of Chemical Technology*, *3rd edn*, **10**, 548, Kirk Olhmer (1978)
3 ANON., *Chemical Age*, 11 (29 October 1971)

CHAPTER 1

Some fundamental aspects of fluidization

1.1 General features of gas–solid fluidization

Suppose a quantity of finely divided solid particles is placed in a cylindrical container fitted with a porous baseplate and air is pumped into the cylinder through the base. If the top of the cylinder is open to the atmosphere the air will percolate through the particles and emerge at the surface. Now if the air pressure at the base of the bed is increased gradually the gas velocity, and hence the drag force acting on the particles, will also increase until a point is eventually reached at which the drag force is equal to the gravitational force holding the particles within the container. At this point the particles will begin to move apart and the bed will expand upwards as the particles become suspended in the flowing air stream. The particles are then said to be 'fluidized' and in this condition the bed takes on the appearance of a liquid having a flat surface and responding in the same way as a liquid to stirring or pouring. The superficial gas velocity at which this occurs is called the 'minimum fluidization velocity', U_{mf}, and its value depends on the physical properties of the gas and of the solid particles.

Now if the gas velocity is increased above U_{mf}, it is observed experimentally for many powders that gas bubbles form within the bed and rise to the surface where they burst in the same way as gas bubbles in a boiling liquid. At these velocities the bed is divided essentially into two phases–the dense or emulsion phase where gas percolates through, rather as in a packed bed, and the lean or bubble phase where much of the gas is out of contact with solids. A further increase in gas velocity leads to more vigorous bubbling with bigger and bigger bubbles appearing at the surface until at some velocity, U_s, the diameter of the surface bubbles becomes equal to the diameter of the containing vessel itself. The bed is then said to be 'slugging' and it is characterized by a considerable heaving of the surface with many particles being thrown up into that space above the bed known as the 'freeboard'.

Still higher gas velocities result in more and more particles being thrown into the freeboard, with many being transported out of the bed altogether, until at the highest velocities only a low concentration of rapidly moving particles remains in contact with the gas. This latter is the region of dilute-phase flow sometimes found in pneumatic conveying systems. For finely divided powders there is evidence for the existence of yet another regime between the regions of slugging and dilute-phase flow; this is the so-called 'fast fluidization' regime characterized by the highly turbulent motion of aggregates or clusters of particles, the volume of which may be some twenty-five per cent of the total bed volume.

Fluidizing a bed of solid particles with a gas provides a means of bringing the two into intimate contact and this can be very useful in an industrial context where large quantities of solid material have to be processed. Furthermore, experiments with beds of fluidized solids have shown them to possess other properties which enhance this general usefulness. Thus even when accommodating strongly exothermic or endothermic processes, fluid beds are normally completely isothermal with the solid particles being very well mixed; also, owing to their liquid-like properties, fluid beds are capable of being transferred mechanically or 'pumped' from one container to another so facilitating the handling of solids which need to be subjected to different physical or chemical operations in different vessels.

There are however some disadvantages to the use of these systems for industrial processes. For example, although particle mixing is generally good, gas mixing is not and the flow of bubbles, which are a permanent feature of beds operated at gas velocities in excess of the minimum fluidization value, can severely reduce the degree of contact between the gas and the bed particles. There can also be the problem of particle break-up or attrition caused by the vigorous agitation of the particles and their impingement on vessel walls, internals, etc. Nevertheless the advantages often outweigh the disadvantages and fluid beds find wide and increasing application in many branches of the chemical and allied industries.

1.2 Minimum fluidization velocity

The gas velocity at which a bed of powder becomes fluidized may be found by measuring the pressure drop across the bed as a function of gas velocity as shown in *Figure 1.1*. It is normal practice to work from the higher velocities at which the bed is vigorously fluidized down to the lower velocities in order to prevent spurious readings caused by particle interactions in the packed-bed state. Many powders behave according to the 'normal' curve shown and begin to bubble as soon as the minimum fluidization velocity, U_{mf}, has been exceeded; for these materials U_{mf} and the minimum bubbling velocity, U_{mb}, are equal. Some powders however show 'abnormal' behaviour and although fluidized, continue to expand uniformly beyond U_{mf}; when U_{mb} is reached the particles collapse and the bed begins to behave more or less normally. The reasons for this difference in behaviour will be discussed in Section 1.3.

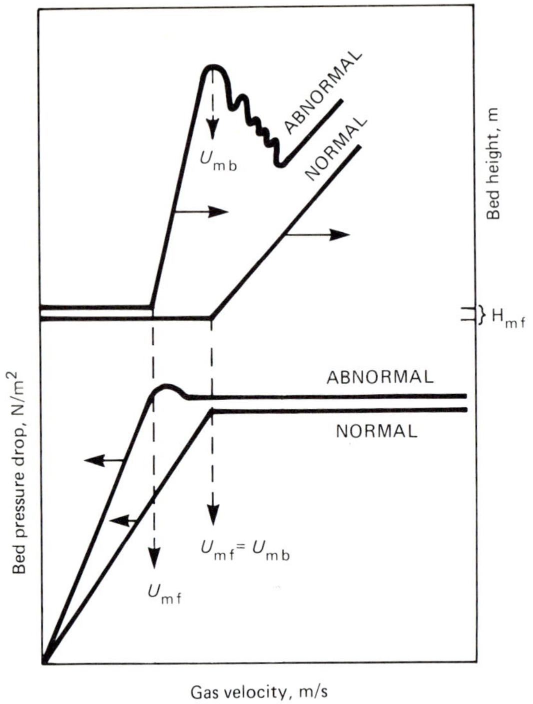

Figure 1.1 Pressure drop through bed as a function of gas velocity in the neighbourhood of minimum fluidization velocity

A number of empirical expressions are available which give the value of U_{mf} for a particular powder in terms of the physical properties of the solid particles and of the fluidizing gas. They are mostly based on the principle of taking a gas velocity–pressure drop relationship for a packed bed of particles and extending it to the point where the particles become fluidized and the gas velocity through the bed is U_{mf}. The velocity of gas moving in laminar flow through a bed of porous material of depth h is given by Darcy's law as:

$$U = C \frac{\Delta p}{h} \tag{1.1}$$

where Δp is the pressure drop through the bed and C, the permeability of the medium, is a function of both fluid viscosity and the fraction of voids in the bed. On the assumption that a bed of particles could be considered equivalent to an assembly of parallel cylindrical channels with internal surface and volume equal to the particle surface and pore volume respectively, Kozeny[1] combined equation 1.1 with the Hagen–Poiseuille equation for laminar flow through cylindrical tubes of diameter d:

$$U = \frac{d^2}{32\mu} \frac{\Delta p}{h} \tag{1.2}$$

and obtained the expression:

$$U = \frac{\epsilon^3}{k\mu S^2} \frac{\Delta p}{h} \tag{1.3}$$

Here ε is the bed void fraction, μ the fluid viscosity, S the particle surface area per unit volume of bed and k is a constant. Now for spherical particles of diameter d_p:

$$S = \frac{6(1 - \varepsilon)}{d_p} \tag{1.4}$$

and so substituting for S in equation 1.3:

$$U = \frac{\varepsilon^3}{(1 - \varepsilon)^2} \frac{d_p^{\,2}}{36k\mu} \frac{\Delta p}{h} \tag{1.5}$$

It was later pointed out by Carman[2] that since the pores in real materials are not straight the actual paths taken by the fluid are sinuous and so the value of the Kozeny constant k should be different from that assumed by Kozeny himself. On the basis of experimental evidence, Carman proposed a value for k of 5.0 thus enabling equation 1.5 to be written:

$$U = \frac{\varepsilon^3}{180(1 - \varepsilon)^2} \frac{d_p^{\,2}}{\mu} \frac{\Delta p}{h} \tag{1.6}$$

This equation is known as the Carman–Kozeny equation and it applies to a wide range of conditions in which fluids flow through porous solids. For beds of non-spherical particles it is necessary to introduce a sphericity factor ϕ defined as:

$$\phi = \frac{\text{surface area of sphere of volume equal to that of particle}}{\text{surface area of particle}}$$

Thus $\phi = 1$ for perfect spheres and $0<\phi<1$ for other shapes. The Carman–Kozeny equation may now be used to obtain an expression for the pressure drop through a vertical bed of particles of height H_{mf}:

$$\frac{\Delta p}{H_{mf}} = \frac{180(1 - \varepsilon)^2}{\varepsilon^3} \frac{\mu U}{(\phi d_p)^2} \tag{1.7}$$

Now, at the point of minimum fluidization the force exerted by the upward flowing fluid is equal to the gravitational force acting on the particles:

$$\frac{\Delta p}{H_{mf}} = (1 - \varepsilon_{mf})(\varrho_s - \varrho_g)g \tag{1.8}$$

where ε_{mf} is the void fraction at minimum fluidization and ϱ_s and ϱ_g are the densities of particles and gas respectively. Substituting the appropriate values of voidage and gas velocity in equation 1.7 and combining it with equation 1.8 gives:

$$(1 - \varepsilon_{mf})(\varrho_s - \varrho_g)g = \frac{180(1 - \varepsilon_{mf})^2}{\varepsilon_{mf}^3} \frac{\mu U_{mf}}{(\phi d_p)^2} \tag{1.9}$$

from which the minimum fluidization velocity is found to be:

$$U_{mf} = \frac{\varepsilon_{mf}^3}{180(1 - \varepsilon_{mf})} \frac{(\varrho_s - \varrho_g)(\phi d_p)^2 g}{\mu} \tag{1.10}$$

The Carman–Kozeny equation is strictly applicable only to conditions of laminar flow where the pressure drop is due solely to viscous energy losses. With beds of particles greater than about 150 μm in diameter inertia forces become important and it is then necessary to use the Ergun equation[3] in the determination of U_{mf}:

$$\frac{\Delta p}{h} = \frac{150(1 - \varepsilon)^2}{\varepsilon^3} \frac{\mu U}{(\phi d_p)^2} + \frac{1.75(1 - \varepsilon)}{\varepsilon^3} \frac{\varrho_g U^2}{\phi d_p} \tag{1.11}$$

One of the problems in applying the Carman–Kozeny or Ergun equation to the calculation of U_{mf} is that the minimum fluidization voidage is frequently unknown. Wen and Yu[4] found that for a range of particle types and sizes the following two empirical relationships were valid:

$$\frac{1 - \varepsilon_{mf}}{\phi^2 \varepsilon_{mf}^3} \simeq 11; \quad \frac{1}{\phi \varepsilon_{mf}^3} \simeq 14 \tag{1.12}$$

They combined these with the Ergun equation and obtained the generalized correlation:

$$Re_{mf} = [(33.7)^2 + 0.0408 Ga]^{1/2} - 33.7 \tag{1.13}$$

$$\text{where } Re_{mf} = \frac{U_{mf} d_p \varrho_g}{\mu} \tag{1.14}$$

$$\text{and } Ga = \frac{d_p^3 \varrho_g (\varrho_s - \varrho_g) g}{\mu^2} \tag{1.15}$$

from which U_{mf} may readily be calculated knowing only the physical properties of gas and solids.

Fluidized beds rarely contain powders with a uniform particle size; more often than not a distribution of sizes is present and it is then necessary to define some convenient average dimension for characterization purposes. The most frequently used average is the surface mean diameter defined as:

$$\bar{d}_{sm} = \frac{1}{\sum \frac{x}{d_a}} \tag{1.16}$$

where x is the mass fraction of particles in each size range given by the sieve aperture d_a; $\bar{d}_{sm}$ is the diameter of a particle with an external surface-to-volume ratio equal to that of the mixture average.

1.3 Fluidization characteristics of powders

Although for many powders the gas velocity-pressure drop relationship is the 'normal' one shown in *Figure 1.1*, by no means all materials behave in the same way. Some undergo considerable bed expansion before the onset of bubbling at U_{mb} ('abnormal' curve on *Figure 1.1*) at which point the bed collapses to the normal value. There are other powders that are extremely difficult to fluidize at all, the gas merely channelling its way through the bed which remains practically static at all velocities up to the very highest. A convenient classification has been suggested by Geldart[5] in which the various powders are arranged into four groups as follows:

Group A powders behave in the abnormal manner indicated above and are characterized by a relatively small particle size (d_p = 30–150 μm) and a low particle density (<1500 kg/m^3). Frequently these powders show an increase in the void fraction of the emulsion phase as the fluidizing gas velocity increases. Furthermore the minimum bubbling velocity U_{mb} is always greater than U_{mf}. They easily form bubbles which rise at velocities much in excess of that of the emulsion phase gas. Fluid cracking catalyst is a typical example of a Group A powder (*see* Chapter 3).

Group B powders behave normally and begin to bubble at gas velocities just in excess of U_{mf}. They are typically materials with d_p = 150–500 μm and densities in the range 1500–4000 kg/m^3. As with Group A materials bubble velocities are high compared to dense phase gas velocities. Their dense phase void fractions remain more or less constant and close to the minimum fluidization value throughout the normal range of gas velocities. Examples of Group B materials are coarse sand and glass beads (ballotini).

Group C. Cohesive powders of very small size (d_p<30 μm) and whose shape is far from spherical (e.g. talc) come into this category. They are difficult to fluidize and easily give rise to 'channelling'.

Group D are fluidizable materials that are either of very large size or very dense. Normally their dense phase gas velocities are high, while any bubbles formed rise rather more slowly than in the case of powders in Groups A or B. Lead shot is an example of a Group D material.

A 'phase diagram' showing the boundaries of the various groups is given in *Figure 1.2*.

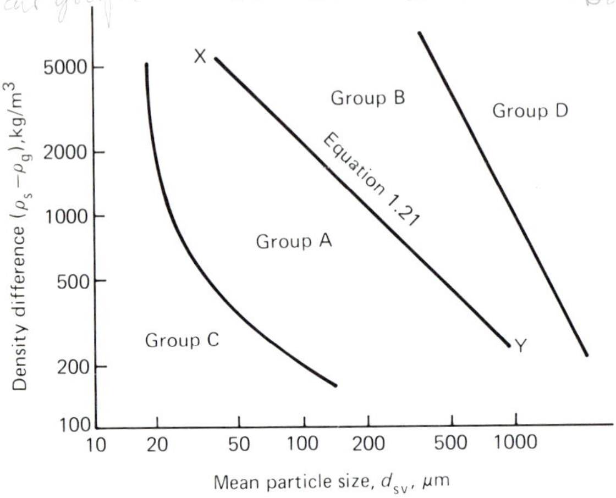

Figure 1.2 Powder classification diagram for fluidization by air–ambient conditions (From Geldart[5])

Geldart[5] measured U_{mf} values for a number of materials of different size and density and correlated the data in terms of Reynolds and Galileo numbers (equations 1.14, 1.15) somewhat after the manner of Wen and Yu[4]:

$$Ga = 1823\, Re_{mf}^{1.07} + 21.7\, Re_{mf}^{2} \qquad (1.17)$$

In the expressions for Re_{mf} and Ga, the surface mean diameter (equation 1.16) of the particles is used. For Group A powders the laminar part of equation was found to be sufficient, and thus:

$$U_{mf} = \frac{9 \times 10^{-4}\, (\varrho_s - \varrho_g)^{0.934} g^{0.934} (\bar{d}_{sm})^{1.8}}{\mu^{0.87} \varrho_g^{0.066}} \qquad (1.18)$$

Values for the minimum bubbling velocity, U_{mb}, were also measured and found to be directly proportional to particle size:

$$U_{mb} = K_{mb}\, \bar{d}_{sm} \qquad (1.19)$$

Equations 1.18 and 1.19 are plotted for two powders of different densities in *Figure 1.3*. For particles with $(\varrho_s - \varrho_g) = 10^3$ kg/m^3 and $\bar{d}_{sm} = 100\,\mu$m, the graph shows a U_{mf} value of 4.1 mm/s and a U_{mb} value of 9.5 mm/s, indicating a region of bubble-free expansion just beyond the point of incipient fluidization. A bed of 100 μm particles with $(\varrho_s - \varrho_g) = 2.7 \times 10^3$ kg/m^3 is shown to have $U_{mf} = 10.5$ mm/s and $U_{mb} = 9.5$ mm/s and hence this Group B powder will bubble at the minimum fluidization velocity.

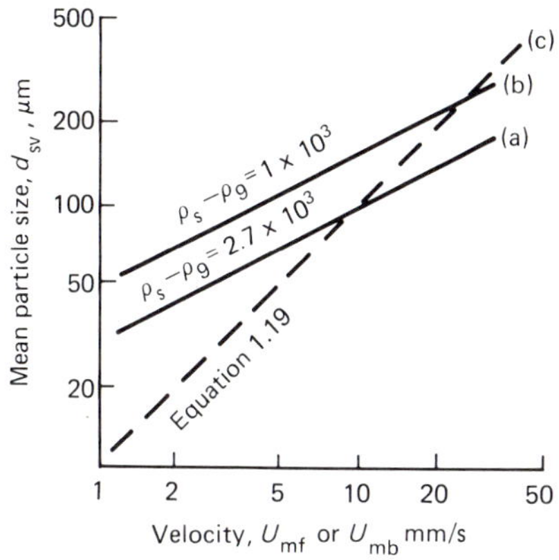

Figure 1.3 Minimum fluidization velocity (a,b) and minimum bubbling velocity (c), for air, v. mean particle size (From Geldart[5])

The boundary between Group A and Group B behaviour is defined as the line on a particle density-diameter plot for which $U_{mf}/U_{mb} = 1$. Then for Group A powders $U_{mf}/U_{mb} < 1$ and for Group B $U_{mf}/U_{mb} \geqslant 1$. From equations 1.18 and 1.19 the line is thus given by:

$$\frac{9 \times 10^{-4} (\varrho_s - \varrho_g)^{0.934} (\bar{d}_{sm})^{0.8} g^{0.934}}{\mu^{0.87} \varrho_g{}^{0.066} K_{mb}} = 1 \qquad (1.20)$$

and for air at ambient conditions as the fluidizing gas:

$$(\bar{d}_{sm})(\varrho_s - \varrho_g)^{1.17} = 906\,000 \qquad (1.21)$$

where $(\bar{d}_{sm})$ is in μm and $(\varrho_s - \varrho_g)$ is in kg/m^3. Equation 1.21 is plotted as *XY* on the 'phase diagram', *Figure 1.2*. The C–A and B–D boundaries are empirical lines.

Geldart's empirical classification has proved extremely useful in predicting the behaviour of various powders at and near the point of minimum fluidization, but there is as yet no convincing explanation of the reason for this behaviour.

1.4 Interparticle forces

It is assumed generally that at the point of minimum fluidization the particles are freely supported on the upward flowing gas stream and that there is then no interaction between them. While this may well be true for beds of relatively large, coarse particles (Geldart Group B) it may not be so for powders such as those of Group A that contain a high proportion of 'fines', i.e. particles with a mean size less than about 45 μm. As we have seen these powders undergo considerable expansion in the velocity region between U_{mf} and U_{mb}, the minimum bubbling value, and bubble formation is accompanied by bed collapse. A number of authors have suggested that this is due to the formation and subsequent loss of a loose structure of cavities among the particles and have proposed both van der Waals forces and capillary condensation forces as being responsible for holding particles together.

Massimilla and Donsi[6] give the following equation for the attractive van der Waals force, F_w, between two non-deformable spheres of radii R_1 and R_2 separated by a distance Z_o:

$$F_w = \frac{h\omega}{8\pi Z_o^2} R \qquad (1.22)$$

$$\text{where } R = \frac{R_1 R_2}{R_1 + R_2}$$

$h\omega$ is the Lifshitz-van der Waals constant which ranges from 1.6×10^{-19} to 16×10^{-19}J (1–10 eV) for most solids in air. For deformable solids the potential increase in area of contact leads to an increase in the cohesive force:

$$F_w = \frac{h\omega}{8\pi Z_o^2} \left[1 + \frac{h\omega}{8\pi^2 Z^3_o H}\right] R \qquad (1.23)$$

where H represents the hardness of the more deformable solid.

The capillary force arising from the formation of a liquid bridge between particles in contact is a function of surface tension, σ, and particle radius[7]:

$$F_c = 8\sigma R \qquad (1.24)$$

Capillary forces can result from condensation of water vapour present in the fluidizing air and they can be an order of magnitude greater than the van der Waals forces. Both can greatly exceed the weight of individual particles.

Hence under favourable conditions particles can cohere causing their behaviour in a fluid bed to be quite different from what would otherwise be expected. It has been shown however that the strong capillary forces are almost entirely absent from systems where powders are fluidized with air at a relative humidity of less than eighty per cent[8] and that the van der Waals forces between particles are weakened by the presence of surface irregularities and asperities[6]. Nevertheless they are considered to be strong enough to cause some cohesion and to stabilize a type of cavity structure in the powder in the region of bed expansion between U_{mf} and U_{mb}. Furthermore it is not inconceivable that elements of this structure can persist well into the bubbling region.

The matter has also been studied by Rietema and Mutsers[9] who have proposed the use of a bed elasticity term E(kg/ms^2) to take account of inter-particle cohesion. This elasticity coefficient may be combined with the normal hydrodynamic parameters of fluidized beds to give an equation analagous to equation 1.11 for the minimum bubbling condition:

$$\frac{\varrho_s^3 g^2 d_p^4}{\mu^2 E_{mb}} = \left[\frac{150(1-\varepsilon_{mb})}{\varepsilon_{mb}^2(3-2\varepsilon_{mb})}\right]^2 \qquad (1.25)$$

where ε_{mb} is the bed voidage and E_{mb} the bed elasticity both at U_{mb}. The term on the left hand side of equation 1.25 is a dimensionless fluidization number N_f. The authors simplified the equation to obtain ε_{mb} as a function of the square of the ratio of gravitational acceleration and fluid viscosity:

$$\varepsilon_{mb} = f_1\left(\frac{g}{\mu}\right)^2 \qquad (1.26)$$

and compared this with the relationship expected on purely hydrodynamic grounds, i.e. in the absence of interparticle attaction:

$$\varepsilon_{mb} = f_2\left(\frac{g}{\mu^2}\right) \qquad (1.27)$$

Rietema and Mutsers then measured ε_{mb} in two-dimensional beds of polypropylene powder and glass ballotini which were set up in a 'human centrifuge'[a] and rotated at a series of high speeds so as to vary g. They found equation 1.26 to show better agreement with experiment than equation 1.27 and so concluded that interparticle forces are important.

Although there would appear to be little question that particle cohesion can influence behaviour, particularly for beds of fine particles, it is doubtful that it can account for all the known peculiarities of these systems. For instance the application of pressure has been shown to cause a considerable increase in bed expansion up to U_{mb}[10] and yet it is difficult to see what effect pressure has on any of the parameters in equation 1.26. It would thus seem that hydrodynamic factors can predominate but that interparticle forces cannot be ruled out entirely.

[a] A device situated at the National Aerospace Medical Centre, Soesterberg, Netherlands, used for subjecting trainee pilots to variations in apparent gravity.

1.5 Liquid–solid fluidization

A liquid may be used in place of a gas to fluidize a bed of solid particles and, as with a gas, the system is characterized by a minimum fluidization velocity at which the drag force exerted on the particles by the upward flowing fluid just balances the force due to gravity. Unlike gas–solid systems however the majority of liquid fluidized beds do not form bubbles as the liquid velocity is increased beyond U_{mf} but expand in a regular manner with the bed voidage, ϵ, increasing as the liquid velocity increases. Furthermore a sharp interface is maintained between the top of the bed of particles and the liquid above it. Such behaviour has been called 'particulate' fluidization to differentiate it from the 'aggregative' fluidization found in gas–solid systems.

The fundamental reasons for the difference in behaviour of gases and liquids are not at all understood but an important factor seems to be the density ratio, ρ_s/ρ_f, between the solid particles and the fluidizing fluid. If this is high, as with most gas-solid combinations, then aggregative behaviour is observed; if the ratio is low, as with most liquid–solid combinations, then the system behaves in the particulate manner. Changing the ρ_s/ρ_f ratio can lead to a change in behaviour irrespective of the phases involved. Thus a bed of tungsten beads fluidized with water ($\rho_s/\rho_f = 19.3$) shows the formation of bubbles and slugs[11], while light powders fluidized with high pressure gases can be expanded considerably before bubbling occurs[12].

Following on from earlier studies[13–16] Richardson and Zaki[17] established an important quantitative relationship that is widely applicable to particulately fluidized beds. They studied both the sedimentation and fluidization of uniformly sized spherical particles greater than 100 μm in diameter with a number of liquids in tubes of different diameters. Most of the work reported by them was in fact concerned with sedimentation and although several solids of different size and density were studied, the fluidization experiments were confined to tubes of 0.062 m and 0.038 m diameter with water as the fluidizing medium. They plotted superficial liquid velocity, U, against bed voidage, ε, and found the linear relationship:

$$\log U = n \log\varepsilon + \log U_i \tag{1.28}$$

where n is the slope of the line and $\log U_i$ is the intercept on the $\log U$ axis corresponding to $\varepsilon = 1$. In the case of sedimentation U_i is approximately equal to the terminal fall velocity of the particles U_t, but in the case of fluidization a term must be included to take account of wall effects. Thus from equation 1.28:

$$\frac{U}{U_i} = \varepsilon^n \tag{1.29}$$

where $U_i = U_t$ for sedimentation

and $\log U_i = \log U_t - \frac{d}{D}$ for fluidization

The exponent n is a function of both Reynolds number and particle-to-bed diameter ratio, $\frac{d}{D}$:

$$
\begin{aligned}
n &= 4.65 + 20 \frac{d}{D} && (Re_t < 0.2) \\
n &= \left(4.4 + 18 \frac{d}{D}\right) Re_t^{-0.03} && (0.2 < Re_t < 1) \\
n &= \left(4.4 + 18 \frac{d}{D}\right) Re_t^{-0.1} && (1 < Re_t < 200) \qquad (1.30) \\
n &= 4.4\, Re_t^{-0.1} && (200 < Re_t < 500) \\
n &= 2.4 && (Re_t > 500)
\end{aligned}
$$

where Re_t is the Reynolds number under terminal free-fall conditions:

$$Re_t = \frac{U_t d \rho_f}{\mu} \qquad (1.31)$$

Equation 1.29 is generally known as the Richardson–Zaki equation and although purely empirical it accounts well for the behaviour of a wide range of liquid–solid systems. It is limited in application however to systems undergoing particulate fluidization and to beds of more or less uniformly sized solids[18].

Although the hydrodynamics of liquid–solid fluidized beds appear to be rather more readily interpretable than those of gas–solid systems, and hence their design for an industrial application to be correspondingly more straightforward, the number of such uses currently to which they are put is relatively small. Therefore we will not consider them further here but will concentrate on the more widely used gas–solid fluidized beds.

1.6 Bubbles

The bubbles formed at gas velocities in excess of U_{mb} present one of the most intriguing features of fluidized beds. Although they have been the subject of many theoretical and experimental studies the reasons for their origin and subsequent behaviour in the bed remain obscure. Bubbles may readily be observed in a 'two-dimensional' bed formed from two parallel-sided transparent plates separated by a distance of about 1 cm and fitted with a porous plate distributor (*Figure 1.4*). They are seen to be shaped somewhat like the spherical-cap bubbles formed by injecting air into water although the close proximity of the bed walls causes them to adopt a more elongated profile. (In three-dimensional beds viewed by X-rays isolated bubbles are observed to be roughly spherical in shape with an indented

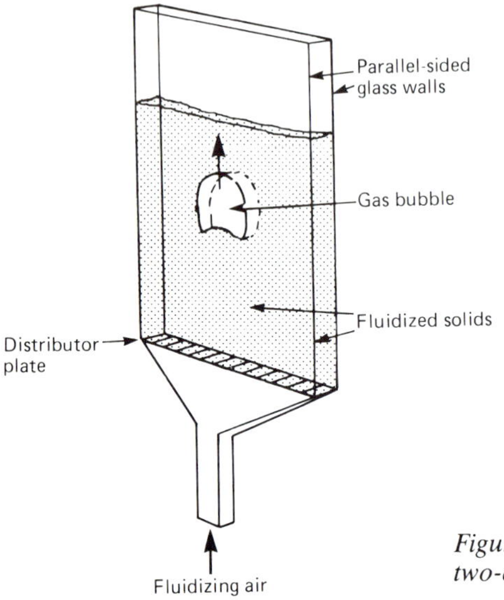

Figure 1.4 Gas bubble rising in a two-dimensional fluid bed

base). Bubbles appear to grow by a process of coalescence as they rise towards the surface and their rise velocity is clearly proportional to their size. They carry a cluster of particles behind them in a wake which fills their indented base, and as they erupt from the bed and the gas they contain escapes into the freeboard, the wake solids are dumped on to the bed surface. It has been show conclusively[19] that it is this transport of solids in bubble wakes that constitutes the prime mechanism by which bed particles are mixed.

Despite some superficial similarities, gas bubbles in fluidized beds are in many ways different from gas bubbles in liquids, and perhaps the most important difference is the absence of a continuous film or 'skin' separating the bubble from the emulsion phase. Because the solid particles in the emulsion are separated from one another by the fluidizing gas stream, the bubble–emulsion interface is not impermeable as it would be in the case of a gas bubble in a liquid, and as a consequence gas can flow out of the bubble into the emulsion; equally of course emulsion-phase gas is not prevented from entering the bubble. This interphase flow can be very important when considering the extent of gas–solid contact in, for example, a fluid-bed catalytic reactor. Clearly the greater the contact the higher will be the chemical conversion that can be achieved. On the other hand bubbles, in some circumstances, can be the cause of serious gas bypassing of the solid particles and this can lead to a lower level of conversion than would otherwise be expected.

Considerations of bubble flow are therefore central to the design of any gas–solid fluid-bed process and so it will be useful to examine various aspects of bubble behaviour in some detail. We will begin with one of the earliest theories to be proposed for the division of gas between the bubble phase and the emulsion phase of a bubbling bed.

1.6.1 Two-phase theory

This theory was proposed originally by Twoomy and Johnstone[20] and states that 'all gas in excess of that necessary to just fluidize the bed passes through in the form of bubbles.' Thus, if Q_T is the total volumetric flowrate into the bed, Q_{mf} the minimum fluidization flowrate and Q_b the bubble flowrate:

$$Q_T = Q_b + Q_{mf} \tag{1.32}$$

Dividing through by the bed area, A, gives the bubble flow in terms of the superficial velocities U and U_{mf}:

$$\frac{Q_b}{A} = U - U_{mf} \tag{1.33}$$

In recent years however a large number of experimental studies have been reported in which the flow of bubble gas in freely bubbling beds has been measured and the general conclusion drawn has been that in the majority of systems the two-phase theory overestimates the visible bubble flow, in some cases by a considerable amount; for a review see Grace and Clift[21]. In other words, in these systems a larger quantity of gas flows through the emulsion phase than is predicted by the theory. It has been suggested therefore that equation 1.33 be recast in the following form:

$$\frac{Q_b}{A} = U - U_{mf}(1 + nf_b) \tag{1.34}$$

where n is a positive number and f_b is the volume fraction of bed occupied by bubbles. In the 'ideal' two-phase theory of course, $n = 0$, and the larger n becomes the larger is the emulsion-phase flow and the greater is the departure from ideality.

A number of workers have obtained values for n on both theoretical and experimental grounds. Thus Lockett *et al.*[22] showed theoretically that for idealized circular (two-dimensional) and spherical (three-dimensional) bubbles the values of n are 1 and 2 respectively. Their treatment was based on the concept of a bubble acting as a low-resistance short circuit for interstitial gas with the through-flow gas velocity being respectively $2U_{mf}$ and $3U_{mf}$ in the circular and spherical bubbles. The gas through-flow idea was also used by Grace and Harrison[23] who found a value of $n = 4.3$ in an experimental study of 274 μm particles of magnesite in a two-dimensional bed. Grace and Clift[21] have shown that the measured deviations from the ideal two-phase theory vary over a wide range and that the value of n not only differs from one system to another but that within a given system it can vary with bed height and gas velocity (*see also* Section 1.6.4).

Bubbles in beds of fine particles normally rise at velocities in excess of the minimum fluidization value and under these circumstances, as will be seen later, a cloud is formed around the bubble. This cloud represents the limit of penetration of bubble gas into the emulsion phase before it is swept back to re-enter the bubble from the rear. In this case therefore through-flow cannot occur, except during coalescence (Section 1.6.4), and deviations from the two-phase theory must be explained in another way. Two theories have been put forward. The first assumes that the velocity of the emulsion phase gas is greater than U_{mf}; this was suggested by Pyle and Harrison[24] who, from observations of visible bubble flow in a two-dimensional bed, concluded that in many cases and particularly in the region close to the distributor, the dense-phase gas velocity was many times greater than the incipient fluidization value. Similar conclusions were reached by Godard and Richardson[25]. The second explanation is that the dense phase can 'expand' and increase its voidage well beyond the value it has at minimum fluidization and that the resulting increase in permeability allows more gas to flow through the emulsion. Persuasive evidence in favour of this second theory has come from the X-rays studies of three-dimensional beds carried out by Rowe and his co-workers. Thus

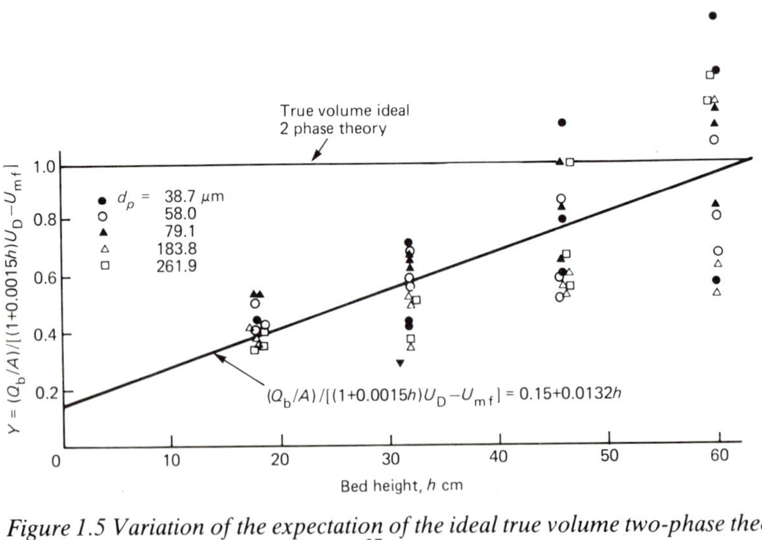

Figure 1.5 Variation of the expectation of the ideal true volume two-phase theory with bed height (From Rowe and Yacono[27])

Rowe and Everett[26] showed that the deficiency in bubble flow found for a range of materials of different particle size could be accounted for by a two percent increase in emulsion phase permeability, while Rowe and Yacono[27] concluded that in their 0.65 m deep beds of closely sized silicon carbide particles, virtually all flow occurred interstitially near the bottom of the bed and gradually decreased to the minimum fluidization value near the surface (*Figure 1.5*). This is in agreement with the remark of Grace and Clift mentioned earlier.

Direct evidence of a change in voidage was obtained by Rowe *et al.*[28] who coupled the X-ray technique with a microdensitometer examination of the dense phase of a bubbling bed of industrial catalyst (a Group A

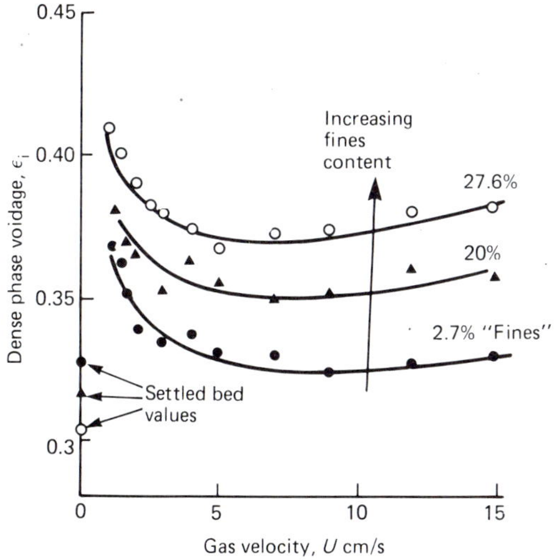

Figure 1.6 Change in dense-phase voidage with gas velocity (From Rowe et al.[28]*)*

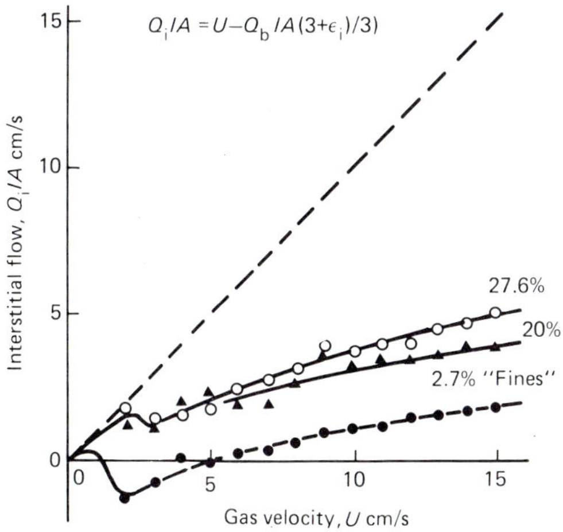

Figure 1.7 Change of interstitial air flow with gas velocity (From Rowe et al.[28]*)*

powder) containing up to some 27% fines ($\bar{d}_p < 45$ μm). They found large changes in voidage with both gas velocity and the fines content of the powder (*Figure 1.6*) and on the basis of a simple model showed that the interstitial flow (*Figure 1.7*) was enormously greater than the minimum fluidization value ($Q_{mf}/A = 2.1$ mm/s). The reason for the strong effect of fine particles on the emulsion phase permeability is not fully understood but is probably linked to the decreases in dense phase kinematic viscosity and bubble stability reported in a number of investigations (*see* Section 1.6.5).

Thus it would appear that the ideal two-phase theory cannot be sustained in beds of either coarse particles, where gas through-flow conditions prevail, or fine particles, where dense-phase expansion occurs, and it would seem that the division of gas flow between the phases of a fluidized bed and what factors determine it are still open to debate.

The influence of the division of gas flow between the two phases on the design of a chemical reactor will be discussed later (Section 2.2.4).

1.6.2 Motion of gas and particles around a bubble

The mechanism of gas and particle flow in bubbling fluidized beds can be investigated by the methods of fluid mechanics and although it is only the most idealized of systems that can currently be treated successfully such studies have done a great deal to aid our understanding of bubble flow and stability. One of the earliest fluid mechanical treatments of bubbles was given by Davidson[29,30] who based his theory on the following assumptions:

(i) the bubble shape is circular in two-dimensional beds and spherical in three-dimensional beds, and the gas pressure inside is constant.
(ii) the bed material behaves like an incompressible fluid of zero viscosity.
(iii) interstitial gas flows like an incompressible viscous fluid that obeys Darcy's law. Davidson then expressed the gas flow in terms of a stream function, ψ_f, defined for a three-dimensional bed by:

$$U_r = -\frac{1}{r^2\sin\theta}\frac{\partial\psi_f}{\partial\theta}$$

$$U_\theta = \frac{1}{r\sin\theta}\frac{\partial\psi_f}{\partial r} \tag{1.35}$$

where U_r and U_θ are components of velocity in the r and θ directions (*Figure 1.8*). He then showed that:

$$\psi_f = (U_b - U_f)\left(1 - \frac{R_c^3}{r^3}\right)\frac{r^2\sin^2\theta}{2} \tag{1.36}$$

where $R_c = R_b\left[\frac{(U_b + 2U_f)}{(U_b - U_f)}\right]^{1/3}$

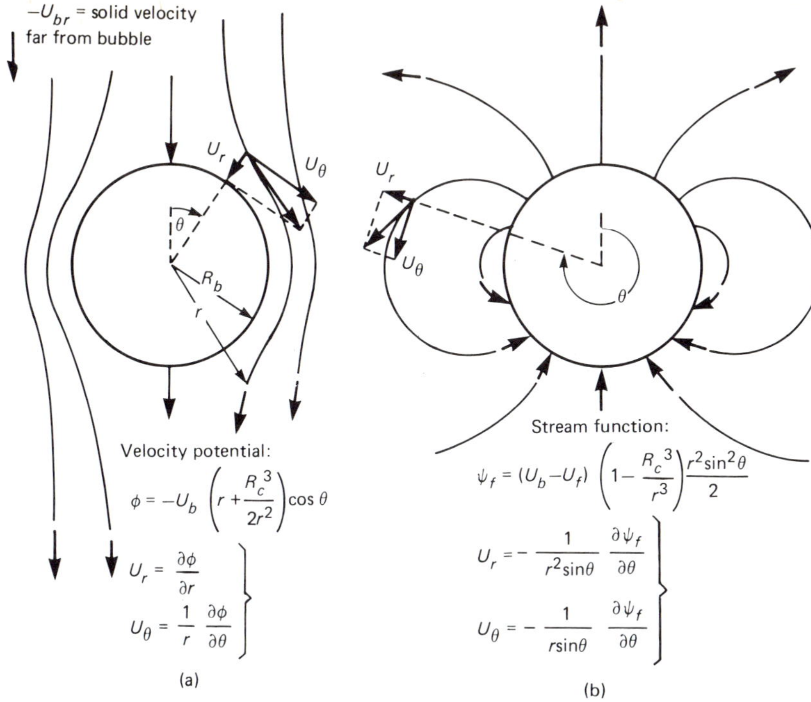

Figure 1.8 Motion of solids in the vicinity of a rising three-dimensional bubble, from potential flow theory; (a) viewed by observer moving with the bubble, (b) viewed by stationary observer (From Kunii and Levenspiel[31])

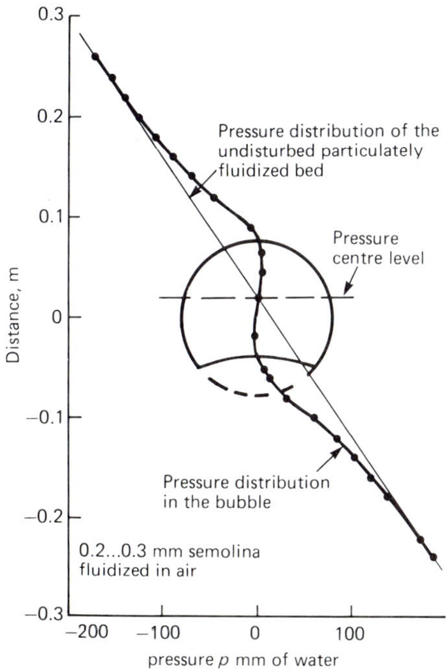

Figure 1.9 Pressure distribution along the axis of symmetry of the bubble (From Reuter[32])

U_b is the bubble velocity, R_b is the bubble radius, and $U_f = U_{mf}/\varepsilon_{mf}$. Davidson showed that the pressure in the upper part of the bubble is higher than that in the adjacent interstitial phase while the pressure in the lower part is lower than that in the neighbouring interstitial phase. This allows gas to flow out of the top under the influence of the pressure gradient and in at the base. Experimental confirmation of this pressure distribution was given by the studies of Reuter[32], *see Figure 1.9*. The bubble is prevented from collapsing by the pressure exerted on the particles by the gas within it although, since there is no 'skin' around the bubble, gas can diffuse out of the void into the particulate phase.

The form of the streamlines is critically dependent on the ratio of bubble velocity to interstitial gas velocity, U_b/U_f. When $U_b<U_f$ (*Figure 1.10(b)*), a situation commonly encountered in beds of large particles, gas flows in at the lower part of the bubble and out at the top using the bubble as a low resistance short circuit in its passage through the dense phase. Some gas however circulates in an annular ring between emulsion and bubble, the

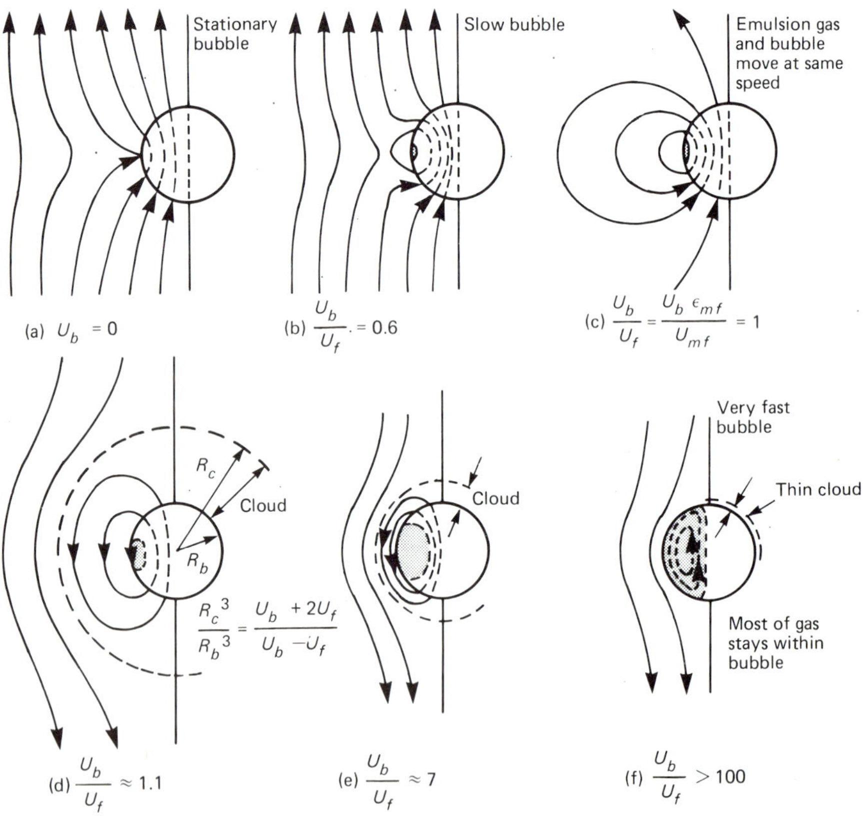

Figure 1.10 Streamlines of gas near a single rising bubble, from the Davidson model. Only flow on the left side is shown; the right side is symmetric (From Kunii and Levenspiel[31])

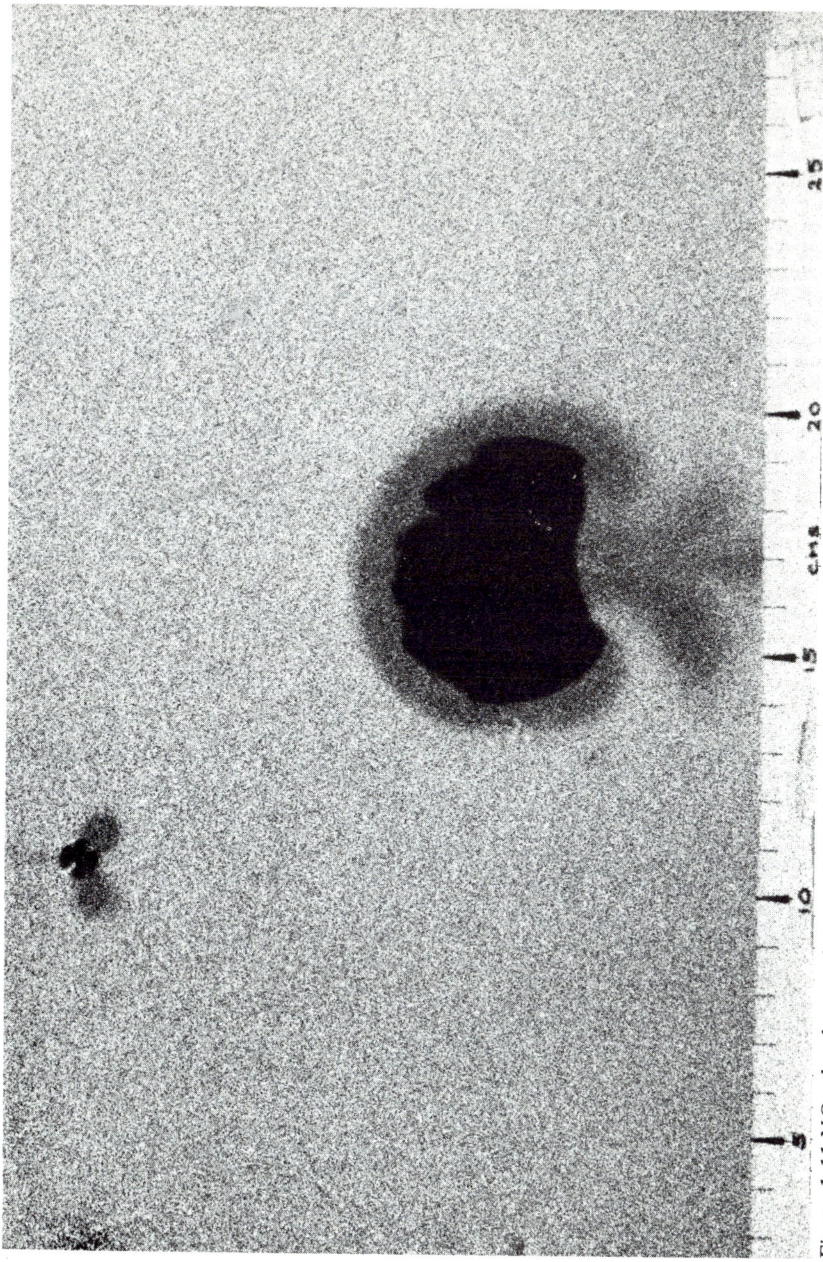

Figure 1.11 NO_2 cloud around a two-dimensional bubble (From Rowe[33])

ring rising with the bubble through the bed. In beds of small particles bubbles normally rise faster than interstitial gas and for this case Davidson's theory predicts that the gas flowing out of the top of the bubble will be dragged back by the downward moving bed particles and forced to re-enter the bubble at its base (*Figure 1.10(d)*). Thus bubble gas is trapped within a region of radius R_c bounded by the stagnant streamline $\psi_f = 0$, and circulates within this volume for as long as the bubble remains in the bed.

The region of penetration of bubble gas into the emulsion is called the 'cloud' region and its existance has been demonstrated by the gas tracer experiments of Rowe[33]. *Figure 1.11* shows a bubble of coloured NO_2 gas injected into a two-dimensional bed at incipient fluidization; the cloud around the upper region is clearly visible. The theory also predicts that as the bubble velocity increases the downward moving particles will cause the circulating gas to be confined ever more closely to the periphery of the bubble and thereby reduce the thickness of the cloud (*Figure 1.10(e) and (f)*). The consequences of cloud formation for fluid-bed processes which rely on good contact between gas and solids can be quite serious since the gas in fast-moving bubbles surrounded by thin clouds could effectively bypass the bed with little or no interaction. Fortunately, as will be seen later, there is in most cases some exchange of gas between the phases and gas–solid contact is not quite as poor as is predicted by theory.

The Davidson theory has provided much insight into bubble flow mechanisms in fluidized beds but it is based on the oversimplified assumption of circular or spherical bubble shapes. Bubbles in three-dimensional beds are normally only roughly spherical with an indented base and carry behind them a wake of particles which can occupy as much as one third of the bubble volume[34]. Theories of gas and particle flow in these more realistic cases have been developed by Jackson[35,36], Murray[37,38] and Collins[39,40] but the relatively simple treatment of Davidson, despite its shortcomings, is still usefully applied in many circumstances.

1.6.3 Velocity

The rise velocity, U_b, of gas bubbles in fluid beds is an important parameter in the design of processes that depend on gas–solid contact. As we have seen, if U_b is much greater than the minimum fluidization velocity, gas bypassing can occur and be a serious problem; if on the other hand $U_b < U_{mf}$ gas–particle contact can be good. Ideally we would like to be able to calculate U_b for a given type of bed material and gas flowrate but unfortunately the problem is quite complicated particularly where freely bubbling beds are concerned and it is doubtful whether there is any generally agreed procedure. For single bubbles of gas rising in beds of liquids of low viscosity Davies and Taylor[41] showed experimentally that:

$$U_b = 0.711\,(gd_b)^{1/2} \tag{1.37}$$

where d_b is the diameter of the sphere with the same volume as the bubble. It was first suggested by Davidson and Harrison[30] that this equation may also be applied to single bubbles of gas rising in fluidized beds and the proposal has gained a wide measure of acceptance. In freely bubbling beds however bubbles undergo coalescence as they rise through the bed, a process which involves the acceleration of a trailing bubble into the rear of a leading one, and which results in the average bubble velocity being higher than that given by the Davies–Taylor equation. Davidson and Harrison also proposed on theoretical grounds that the average velocity of a bubble in a swarm should be given by the natural rising velocity of an isolated bubble plus the upward velocity of the particulate phase between the bubbles:

$$\bar{U}_b = (U - U_{mf}) + 0.711\,(gd_b)^{1/2} \tag{1.38}$$

Although the validity of this equation is doubtful[42] it is widely used in reactor design calculations[31]. It does presuppose of course that a value for d_b is available (*see* Section 1.6.4).

A large number of experimental investigations of bubble rise velocities have been reported[43–47] but there seems to be little general agreement among them. One problem is the different behaviour found in two- and three-dimensional beds with materials that may have been drawn from any of the Groups A–D of the Geldart classification, and it is questionable whether any single equation will correlate data across so wide a range. It is however worth mentioning an equation for bubble velocity that was found to represent quite well data taken from an analysis of some 12 000 bubbles in three-dimensional beds of narrow cuts of particles in the range 40–260 μm[27]:

$$\bar{U}_b = (U - U_{mf}) + 0.0936k_v\,(h + h_0)^{3/8}\,(U - U_{mf})^{1/4}\ \text{m/s} \tag{1.39}$$

The coefficient k_v increases from about 1.0 for 260 μm particles to 1.3 for those of 40 μm, and h_0 is a constant characteristic of the distributor, being approximately zero for a porous plate. An advantage of this equation is that it requires information only on particle properties and distributor design.

The rise velocity of bubbles is also influenced by their shape–ovary ellipsoidal bubbles rising faster than spherical-cap bubbles of the same volume[48]. This has important consequences if vertical rods or tubes (cyclone dip legs, heat exchangers, sampling probes, etc.) are present in a bed since gas bubbles tend to be stabilized on such structures and to become elongated. This results in a reduction in their residence time in the bed although the effect is to some extent compensated by the reduced degree of coalescence caused by the stabilizing influence of the rods.

1.6.4 Growth and coalescence

That bubbles grow by coalescing together as they rise through a fluidized bed is immediately obvious from a superficial examination of a bubbling two-dimensional bed. The region of the bed near the distributor is seen to contain large numbers of small bubbles, whereas towards the bed surface there are a small number of relatively large bubbles. It is widely accepted that coalescence occurs by a trailing bubble accelerating into the rear of a leading one and that the resulting bubble volume is always greater than that of the uncoalesced pair. Thus Grace and Venta[49] from an analysis of cine film of coalescence in a two-dimensional bed concluded that growth in volume occurs both while the leading bubble is being overtaken and while the composite bubble is achieving its final shape. For out-of-line coalescence the increase in total projected area of the two-dimensional bubbles following coalescence was typically some fifty per cent. It is likely that at least some of the deviation from the two-phase theory mentioned earlier can be accounted for on the basis of gas throughflow between bubbles during this process[21]. Thus the observation (*Figure 1.5*) that visible bubble flow near the distributor is less than that predicted by the two-phase theory could be explained by the fact that in this region of the bed coalescence is greatest and so at any time a large amount of gas is 'in transit' between coalescing bubbles. This gas is of course 'invisible' but its presence between bubbles must cause the emulsion phase in that region to expand to a greater voidage than it would normally have in the absence of coalescence. Towards the top of the bed however where coalescence is much reduced there will be less throughflow between bubbles and the visible bubble gas will approximate more closely to the amount predicted by the two-phase theory.

Some of the complexities involved in coalescence were revealed in a study by Sit and Grace[50] who investigated the transfer of ozone gas between pairs of bubbles and the emulsion phase in a two-dimensional bed of glass ballotini. They showed that the process could be divided into four distinct periods: (i) the approach period from the time both bubbles had fully formed to the moment when the nose of the overtaking bubble reached the wake boundary of the leader, (ii) the encroachment period during which the nose of the trailing bubble passed through the wake region and touched the lower surface of the leader, (iii) the coalescing period when although the bubbles had merged they had not attained their final shape, (iv) the post coalescence period starting from the moment the coalesced bubble had taken up its final shape. It was found that the concentration profile of ozone in the bubbles changed most rapidly during the approach and encroachment periods, 77 per cent of the transfer occuring within the first 50 per cent of the timespan covered. Furthermore greater transfer occurred from the leading bubble than was observed from either the trailer or a single bubble of comparable size. The mass transfer coefficient experienced by the leading bubble of a coalescing pair was estimated to be two to three times greater during the approach period than when the bubble was in isolation. The implications of this for reactor design are obvioulsy important (Chapter 2).

The physical reasons for the growth of bubbles by coalescence are as yet unknown. A number of experimental studies have however been carried out to obtain correlations between the operating parameters and the bubble size variation through the bed. These studies have used a variety of techniques to observe the bubbles in three-dimensional beds: surface photography by Whitehead and Young[51] and by Geldart[52,53], capacitance probes by Werther[54] and by Masson and Jottrand[55], optical probes by Yasui and Johanson[56] and X-ray photography by Rowe and co-workers[57,58,59].

A mechanistic model of bubble coalescence was proposed by Darton *et al.*[60] which leads to an equation predicting bubble diameter closely similar to those resulting from empirical studies. The model is based on the assumption that bubbles tend to rise in vertical streams from the distributor but the lateral movement occurs during coalescence between neighbouring streams, the distance travelled by two bubbles before coalescence being

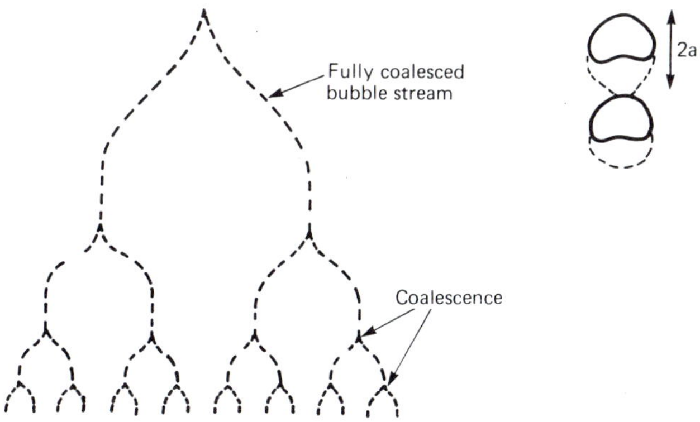

Figure 1.12 Bubble paths during coalescence. The process should be visualized in three dimensions (From Darton et al.[60])

proportional to their horizontal separation (*Figure 1.12*). As already discussed, coalescence normally takes place by a trailing bubble merging into the rear of a leading one and it is assumed in the model that the relatively slow lateral movement which this involves determines the overall rate of growth. From the work of Harrison and Leung[61] the average separation between bubbles is taken to be twice the radius of the spherical region enclosing the bubble and its wake. Assuming the two-phase theory to apply, the variation of bubble diameter with height is then calculated to be:

$$d_b = \frac{0.54\,(U - U_{mf})^{0.4}\,(h + 4\sqrt{A_o})^{0.8}}{g^{0.2}} \tag{1.40}$$

where d_b is the diameter of the sphere with the same volume as that of the bubble, h is the height above the distributor, g is the acceleration due to gravity and A_o is the so-called 'catchment area' which is taken to be the area of distributor per orifice; A_o is considered to be zero for a porous plate distributor. The units in equation 1.40 are SI. Although this is a mechanistic model the expression for d_b contains an empirical constant, 0.54, derived from the work of Werther[54] and Rowe and Everett[58] but the agreement this produces with a wide range of experiemental observations is nevertheless quite good. The model is however inapplicable to slugging beds and to beds containing a high concentration of fine particles in which there is evidence that bubbles reach a stable, limiting size.

1.6.5 Stability

Matsen[62] fluidized a bed of petroleum coke containing 50 per cent by weight of particles below 60 μm in diameter (Geldart Group A) in a 0.60 m (2 ft) diameter column. He measured bed expansion and average density, ρ, and found both to vary linearly with fluidizing gas velocity. Now from equations 1.33 and 1.38 it can readily be shown that:

$$\frac{\rho_{mf}}{\rho} = \frac{H}{H_{mf}} = 1 + \frac{U - U_{mf}}{0.711\,(gd_b)^{1/2}} \tag{1.41}$$

which indicates that bed expansion and bed density can only be linear functions of U if the bubble diameter d_b remains constant. This was taken as experimental evidence in favour of the theory of a maximum stable bubble size first put forward by Harrison *et al.*[63] and later developed by Davidson and Harrison[30]. The theory is based on the assumption that as a bubble rises in a fluid bed the shear force exerted by the particles moving downward relative to the bubble sets up a circulation of gas within the bubble and that the velocity of this circulation, U_c, approximates to the bubble rise velocity, U_b. When, through coalescence, d_b increases to the point where $U_b > U_t$, the terminal fall velocity of the bed particles, the solids in the wake will be drawn up into the bubble causing it to break up into smaller bubbles with lower rise velocities. In this way the bubble would not be expected to exceed a certain size determined by the value of U_t, and beds of small, light particles should therefore contain only relatively small bubbles and show 'smoother' fluidization characteristics. The latter is often true and is frequently remarked upon[64], although experimental evidence in favour of this mechanism of bubble break-up is slight.

An alternative theory, developed by Upson and Pyle[65] and Clift *et al.*[66] following earlier studies by Rice and Wilhelm[67] and Bellman and Pennington[68], suggests that bubble break-up is caused primarily by an instability in the bubble roof allowing particles to rain down through the void and so divide it in two. Curtain formation or 'fingering' is frequently seen in two-dimensional beds although it is not so common in three-dimensional beds observed by X-rays[69]. Instabilities of this type were discussed by Taylor[70] when considering systems in which a heavy fluid is suspended above a lighter one. Upson and Pyle and Clift *et al.* analysed the stability of the bubble surface (assumed for simplicity to be horizontal and two-dimensional) when subjected to a small wave-like disturbance that grows exponentially with time according to exp(nt). They found that all such disturbances produce instability and that there exists a 'most dangerous wavelength' at which the growth rate of the disturbance is a maximum. Whether a bubble splits or not depends on the relation between the rate at which the disturbance grows and the rise velocity of the bubble; thus splitting will occur only if the growing curtain of particles reaches the floor of the bubble before the top of the curtain is swept around beyond the equator. Both analyses are based on a grossly oversimplified description of the bubble surface, since no complete description is yet available, and so their conclusions can give only qualitative indications of the factors that determine bubble stability. They are nonetheless very useful at the present stage of development of the subject.

From the analysis of Clift *et al.* three equations are obtained from which the growth factor n may be evaluated for a gas–solid fluidized bed.

$$R^4 + 2R^2 - 4R + \left(1 - \frac{g}{k^3 \upsilon_p^2}\right) = 0 \qquad (1.42)$$

where k is the wave number of the disturbance,

$$R = \left(1 + \frac{n}{k^2 \upsilon_p}\right)^{1/2} \qquad (1.43)$$

and υ_p, the apparent kinematic viscosity of the emulsion phase, is given by:

$$\upsilon_p = \frac{\mu_p}{\rho_p (1 - \varepsilon)} \qquad (1.44)$$

On the basis of these equations it is clear that the only factor determining the stability of the bubble surface is the apparent kinematic viscosity of the emulsion phase, υ_p, values of which may be obtained by the method of Grace to be described later in section 1.11. Predictions based on the equations are shown in *Figure 1.13* from which it is apparent that an increase in the emulsion phase viscosity results in a decrease in the growth factor and in the 'most dangerous' wave number. In other words the theory predicts an increase in bubble stability as υ_p increases. Measurements by Schügerl[71] of viscosities of fluid beds indicated an increase in υ_p with particle density, an observation also in accord with the theory since

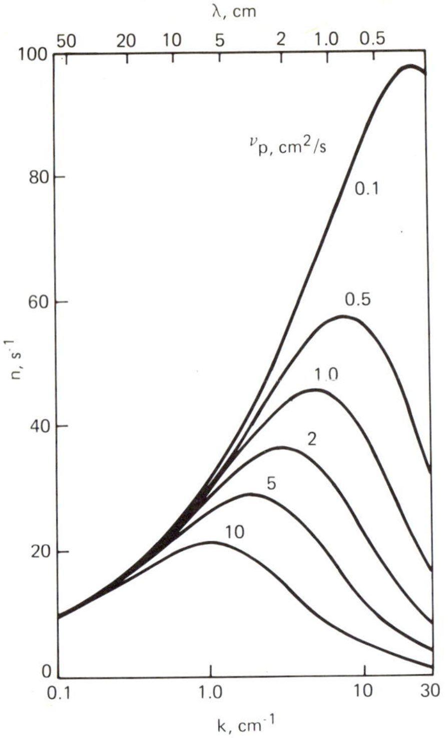

Figure 1.13 Growth rates predicted by simplified equations 1.42–1.44, neglecting density and viscosity of interstitial fluid (From Clift et al.[66])

bubbles would be expected to be stable to larger sizes in beds of higher viscosity and it is widely accepted that beds behave in a more aggregative manner as ρ_s/ρ_f increases. The linearized stability analysis of Upson and Pyle and Clift *et al.* also enables tentative predictions to be made about the bubbling properties of liquid–solid systems. Here, although the dynamic viscosity of the fluidizing medium is higher than in a gas–solid bed, its density is many times greater and so the kinematic viscosity is generally lower. This leads to the conclusion that the growth factor of a disturbance at the surface of a liquid 'bubble' will be high and so the bubble will be unstable and prone to splitting and disintegration.

The analysis also predicts that bubbles should split more readily in fine particle systems (d_p<100 μm) a result consistent with the experimental findings of Matsen already referred to. Fine particle systems however are known to behave in rather curious ways and to show properties different from those containing predominantly large particles. For example, the X-ray study, already referred to[28], of a bed of industrial catalyst in which the fines content was varied, showed that the relative amount of gas flowing interstitially increased as the fines (d_p<45 μm) content was raised from 0 to 27.6 per cent by weight, and furthermore that, for the powder containing the highest proportion of fines, the emulsion phase flow was up to twenty five times greater than the minimum fluidization value. If this is generally true, and evidence would seem to indicate that it is, then fine particle systems contain small bubbles, not because larger bubbles are inherently unstable, but because the emulsion phase can accommodate a greater flow of gas. There is still controversy surrounding this question however, since there is currently no satisfactory hydrodynamic theory to account for the action of fines in promoting interstitial flow at the expense of bubble flow.

1.6.6 Bubbles in beds of large particles

The types of bubble with which we have been concerned so far are those found in fluid-bed systems of relatively small particles (Geldart Groups A and B) and it is these systems that historically have been of greatest industrial interest. Beds of large particles are of growing importance however e.g. in the field of fluidized-bed combustion, and although bubbles do form in such beds their properties are in many ways different from those found in beds of small parricles. Thus because the minimum fluidization velocities of large particles tend to be high the velocity of emulsion phase gas is also high and generally exceeds the bubble rise velocity. Under these conditions no cloud is formed round the bubble and gas simply flows through the void from bottom to top under the influence of the prevailing pressure gradient. As a result there is no gas bypassing in large particle systems, all gas coming into contact with solids in its passage through the bed. Unfortunately relatively little experimental work with large particles has been reported and much of what has appeared in the literature has been concerned with two-dimensional beds where wall effects are even more pronounced than in beds of small particles.

Geldart and Cranfield studied the fluidization characteristics of large particles (d_p>1 mm) of alkalized alumina in both two- and three-dimensional beds[72,73]. They found both U_{mf} and ε_{mf} to vary with bed height in the 2-D bed but to be practically constant in the 3-D case. In both beds however the minimum bubbling velocity remained constant as bed height was changed. Furthermore the visible bubble flow (estimated from cine photographs of the bed surface) was greater in the 3-D bed than in the 2-D bed at corresponding values of $U - U_{mb}$ but was only some 55 per cent of the excess gas flow. They also commented that bubble growth appeared to occur by horizontal rather than vertical coalescence, and that although the rise velocities of single bubbles in the 3-D bed were well represented by the Davies–Taylor equation, in swarms their velocities were reduced owing to the horizintal coalescence interactions.

Few other studies of the hydrodynamics of three-dimensional beds of large particles have been published although the growing importance of these systems is acknowledged in the work of Park *et al.*[74] on atmospheric pressure fluid-bed combustors and of Rios *et al.*[74] on the use of fluid beds in the treatment of large grains of foodstuffs (Sections 4.4.4 and 1.10.1). A considerable body of work exists however on the behaviour of large particles in spouted beds[76], but consideration of these systems is outside the scope of the present study.

1.7 Slugging

In deep fluidized beds of high aspect ratio the continuation of bubble coalescence leads eventually to the formation of bubbles whose diameter is equal to that of the bed itself. Such bubbles are called slugs and their hydrodynamic properties differ in many respects from those of freely moving bubbles; slugs for example rise more slowly than bubbles of equal volume. Furthermore, different types of slug can be formed depending on the type of bed material being fluidized, and *Figure 1.14* shows some of

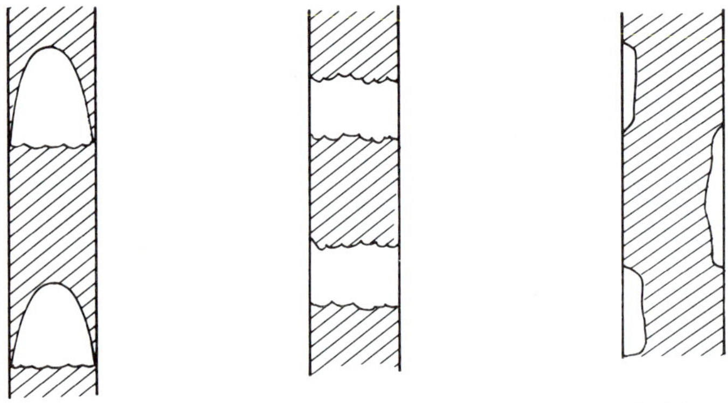

Figure 1.14 Types of slug flow in gas-fluidized beds

those commonly observed[77]. Axisymmetric slugs are normally formed in beds of Geldart Group A powders, i.e. those with finely divided particles of low density. It is worth noting that the same axisymmetric shape is adopted by slugs of gas rising in columns of liquid. In the case of the second type of slug shown in *Figure 1.14*, the leading edge is very flat and this type is more often formed in beds of large, high-density particles. The third type, the asymmetric wall slugs, are usually seen only at high gas velocities and where the flat slugs have broken down[77].

Stewart and Davidson[78] were the first to give criteria for the onset of slugging. They assumed the volume of an axisymmetric slug to be equal to that of a paraboloid of revolution whose height is equal to the diameter of the tube, D. The slug volume is thus $\pi D^3/8$ and the volume fraction of slugs in the bed, f_s, assuming them to be spaced three tube diameters apart, is:

$$f_s = \frac{\pi D^3/8}{3D \times \pi D^2/4} = \frac{1}{6} \tag{1.45}$$

Stewart and Davidson measured slug velocities, U_s, experimentally for a number of powders and found the general relationship:

$$U_s = 0.35\,(gD)^{1/2} + (U - U_{mf}) \tag{1.46}$$

Hence the total slug flowrate, Q_s, will be given by:

$$Q_s = \frac{A}{6}\,[0.35\,(gD)^{1/2} + (U - U_{mf})] \tag{1.47}$$

where A is the cross-sectional area of the bed, and assuming the two-phase theory to hold (i.e. all gas in excess of the minimum fluidization value flows as slugs):

$$Q_s = A(U - U_{mf}) \tag{1.48}$$

By combining equations 1.47 and 1.48 and rearranging, the authors obtained the following equation giving the minimum excess gas velocity at which slugs will form in a tube of diameter D:

$$(U - U_{mf}) = 0.07\,(gD)^{1/2} \tag{1.49}$$

Baeyens and Geldart[79] pointed out however that equation 1.49 implies an absence of effect due to particle size, size distribution and bed height, and while the first two were indeed found experimentally to exert no influence on slug formation, these two authors showed conclusively that bed height is an important variable up to a critical height H_L. They found for a wide range of powders and a number of different sized beds:

$$H_L = 60D^{0.175} \tag{1.50}$$

(where H_L and D are in centimetre units); they also showed that the excess gas flow at which slugging begins in beds of height less than H_L is given by:

$$(U - U_{mf}) = 0.07(gD)^{1/2} + 1.6 \times 10^{-3}\,(H_L - H_{mf})\ \text{cm/s} \tag{1.51}$$

For beds deeper than H_L the criterion of Stewart and Davidson (equation 1.49) holds good. Finally using a method suggested by Matsen and Tarmy[80], Baeyens and Geldart derived an expression for slug frequency, n_s, in terms of bed diameter:

$$n_s = \frac{0.35}{k}\left(\frac{g}{D}\right)^{1/2} \tag{1.52}$$

where k is an empirically determined constant, and showed that n_s tends to a limiting value n_L as bed height exceeds about 1.2 m (for beds up to 0.308 m diameter). They found:

$$n_L = 1.17\ D^{-0.143} \tag{1.53}$$

and so were able to calculate values of k for beds deeper than H_L according to:

$$k_L = 9.38\ D^{-0.357} \tag{1.54}$$

(In equations 1.50–1.54, D is in centimetre units.)

A proper understanding of the hydrodynamics of slugging is important from the point of view of chemical reactor design since many laboratory-scale reactors operate in the slugging mode. Without applying a realistic model of slugging to such reactors it is extremely hazardous to scale up from them to a larger unit where it is likely that freely bubbling conditions will prevail and where the gas–solid contacting pattern will be quite different. This lack of realization of the fundamental differences between slugging beds and bubbling beds has in the past been responsible for a number of costly plant failures[81], but the improved understanding of slugging achieved in recent years now makes it possible to scale up with greater confidence[82].

1.8 Beyond the slugging region

There is some evidence that at gas velocities in excess of those at which slugging occurs, other definable regions of fluidization come into existence. *Figure 1.15* shows a 'fluidization map' given by Yerushalmi *et al.*[83] based on measurements made with cracking catalyst ($U_{mf} \simeq 0.01$ m/s) in a 0.15 m diameter bed. These workers observed a fairly sharp transition from slug flow to turbulent flow at a gas velocity of about 0.6 m/s, while at even higher velocities (>2.2 m/s) a new regime, the so called 'fast' fluidization regime, was established. This latter was characterized by a high degree of particle turbulence with slip velocities, U_{sl}, an order of magnitude greater than the terminal fall velocity of the individual particles. The slip velocity is defined as:

$$U_{sl} = \frac{U_g}{\varepsilon} - U_p \tag{1.55}$$

where U_g is the superficial gas velocity, ε is the bed voidage and U_p the particle velocity. This phenomenon is feasible on hydrodynamic grounds only if the individual particles form into relatively stable agglomerates with correspondingly larger terminal fall velocities. Yerushalmi *et al.*[84] have in fact reported the formation of such clusters of particles at these high

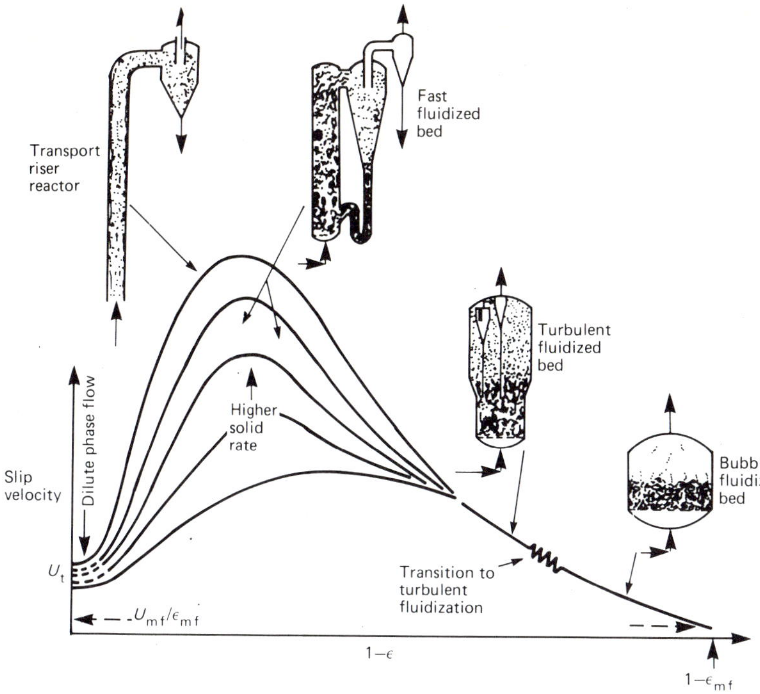

Figure 1.15 Qualitative fluidization map for fine solids (From Yerushalmi et al.[83])

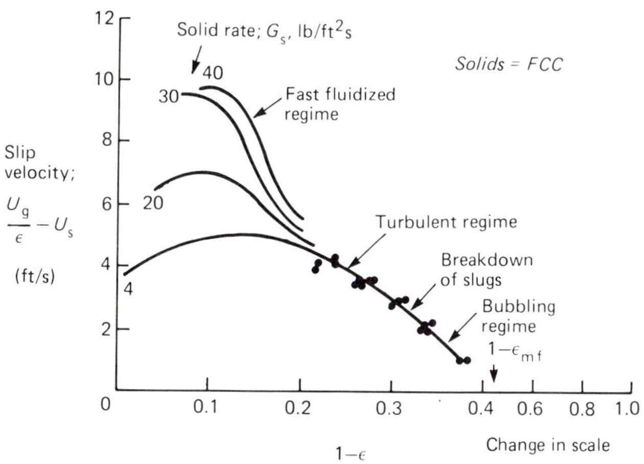

Figure 1.16 A portion of the fluidization map for FCC in 0.15 m bed (From Yerushalmi et al.[83])

velocities and have observed them to undergo considerable back mixing. *Figure 1.16* shows the slip velocities measured relative to the terminal fall velocity of a 55 μm diameter catalyst particle (0.075 m/s) as a function of the solids flowrate in the 0.15 m diameter bed[83]. The equivalent diameter, D_c, of the aggregates formed under these conditions was calculated as follows. Neglecting acceleration and friction effects and assuming the density of the aggregates to be $\rho_s(1-\varepsilon_{mf})$, the pressure drop due to gravity in the vertical pipe is given by:

$$\Delta p = (1-\varepsilon)\rho_s g \Delta L \tag{1.56}$$

$$= (1-\bar{\varepsilon})\rho_s(1-\varepsilon_{mf}) g \Delta L \tag{1.57}$$

where $\bar{\varepsilon}$ is an effective voidage, ρ_s is the solids density and $\Delta p/\Delta L$ is the pressure gradient. The pressure drop due to the drag exerted by the gas on the aggregates is:

$$\Delta p = \frac{N}{A} \frac{1}{2} \rho_g \left(\frac{U_g}{\bar{\varepsilon}} - U_p\right)^2 C_{Dp} f(\bar{\varepsilon}) \frac{\pi}{4} D_c^2 \tag{1.58}$$

where N is the number of aggregates in the length ΔL, A is the bed area, ρ_g is the gas density and U_p is the mean velocity of the aggregates. C_{Dp} is a drag coefficient related to that for a single particle, C_{Ds}, by:

$$C_{Dp} = C_{Ds} \left[\frac{\rho_g D_c \left(\frac{U_g}{\bar{\varepsilon}} - U_p\right)}{\mu} \right] \tag{1.59}$$

μ being the gas viscosity. The function $f(\bar{\varepsilon})$ accounts for the hindered settling caused by the proximity of the aggregates and is related to the Richardson–Zaki index, n, by:

$$f(\bar{\varepsilon}) = \bar{\varepsilon}^{-2n} \tag{1.60}$$

where n is the following function of Reynolds number:

$$n = \begin{cases} 4.65 & Re_t < 0.2 \\ 4.4\, Re_t^{-0.03} & 0.2 < Re_t < 1 \\ 4.4\, Re_t^{-0.1} & 1 < Re_t < 500 \\ 2.4 & Re_t > 500 \end{cases} \tag{1.61}$$

and $Re_t = \rho_g D_c U_t / \mu$.

Hence from equations 1.57, 1.58 and 1.60:

$$D_c = \frac{\left(\frac{3}{4}\right) \rho_g \left(\frac{U_g}{\bar{\varepsilon}} - U_p\right)^2 C_{Dp} \bar{\varepsilon}^{-2n}}{\rho_s (1 - \varepsilon_{mf}) g} \qquad (1.62)$$

Figure 1.17 shows D_c plotted as a function of the solids concentration $(1 - \varepsilon)$ for the 0.15 m diameter pipe.

At still higher velocities than those characterizing the 'fast' regime, the flow in fluidized beds passes into a dilute-phase region where, it is generally agreed, the slip velocity approaches the terminal-fall velocity of individual particles. This type of flow finds its main application in pneumatic conveying.

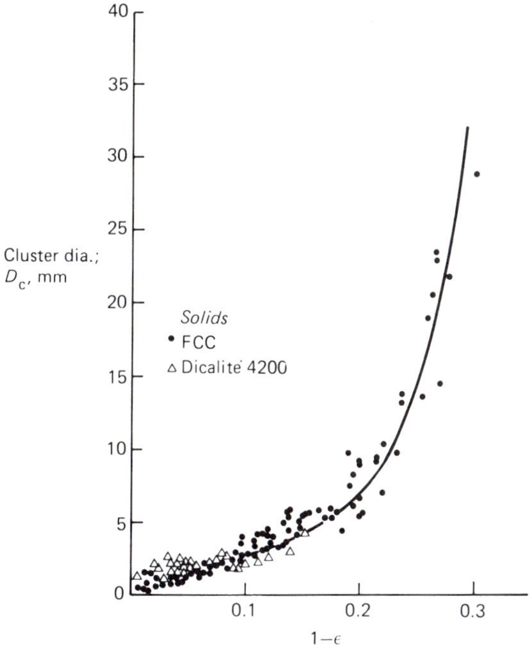

Figure 1.17 Equivalent cluster diameter D_c v. volume concentration of solid (1−ε), in the 0.15 m fast bed (From Yerushalmi et al.[83])

1.9 Entrainment and elutriation

In most practical applications, fluidized beds are operated at gas velocities well in excess of U_{mf} and under these circumstances it is common for particles to be blown from the bed into the freeboard space above it. This phenomenon is known as 'entrainment' and it becomes more severe as the fluidizing gas velocity, U, increases. If U becomes so high that it equals the terminal-fall velocity of the bed particles, U_t, then particles will be 'elutriated', i.e. carried out of the system completely. Normally a fluid bed consists of solid particles with a range of sizes, and as the value of U_t increases with particle diameter the smallest particles are the ones that are elutriated first although the larger ones can be entrained for a certain distance above the surface before falling back to it. Thus it is common for a particle concentration gradient to exist above the bed surface (*Figure 1.18*); the height at which the particle concentration reaches a constant

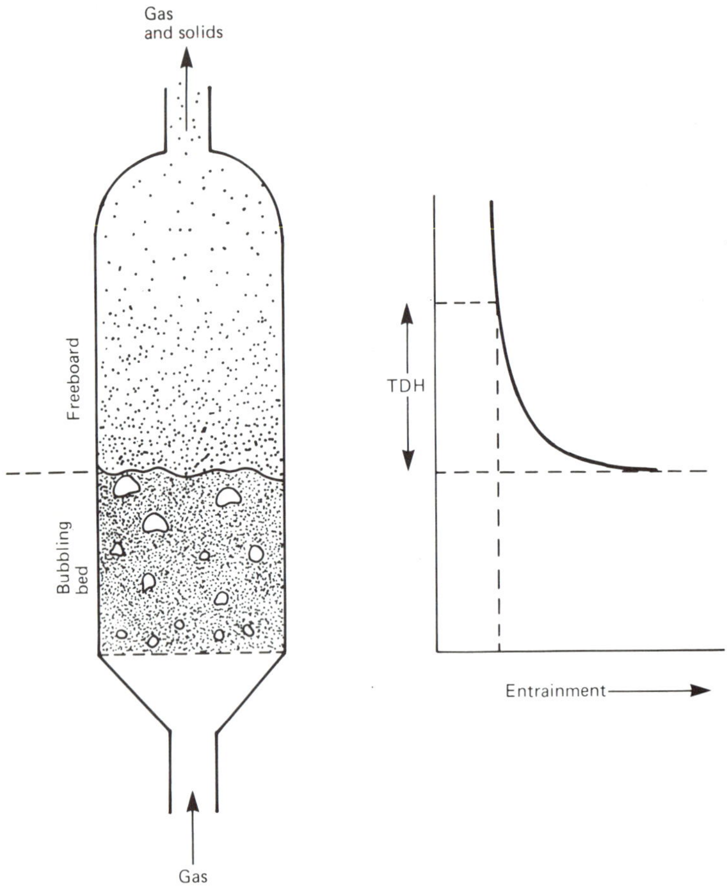

Figure 1.18 Entrainment and elutriation (Based on Bakker and Heertjes[85])

value is known as the 'transport disengaging height' or TDH. Zenz and Weil[86] interpreted the TDH as being that height above the bed surface at which the high-velocity gas jets produced by the erupting bubbles have become dissipated and the gas velocity has returned to U. The particles elutriated from the system are then only those whose terminal fall velocity is less than U, all other particles falling back to the surface.

The terminal velocity, U_t, of a solid particle falling through a fluid may be found from:

$$U_t = \frac{4gd_p(\rho_s - \rho_g)^{1/2}}{3\rho_g C_D} \tag{1.63}$$

In this expression the drag coefficient C_D is itself a function of Reynolds number and so it is necessary to specify the operating range of fluid velocity before U_t can be found. If a particle Reynolds number Re_t is defined as:

$$Re_t = \frac{U_t d_p \rho_g}{\mu}, \tag{1.64}$$

then for spherical particles:

$$\left.\begin{aligned} C_D &= \frac{24}{Re_t} \quad \text{for } Re_t<0.4 \\ C_D &= \frac{10}{Re_t^{1/2}} \quad \text{for } 0.4<Re_t<500 \\ C_D &= 0.43 \quad \text{for } 500<Re_t<200\,000 \end{aligned}\right\} \tag{1.65}$$

Alternatively the velocity independent group:

$$C_D Re_t^2 = \frac{4d_p^3\, g\rho_g\, (\rho_s - \rho_g)}{3\mu^2} \tag{1.66}$$

may be calculated from the known physical properties of solid and fluid and Re_t found directly from empirical correlations[87,88]. The terminal velocity can then be calculated using the expression for Re_t.

Pinchbeck and Popper[89] determined the ratio U_t/U_{mf} for spherical particles over a range of Reynolds numbers and found, as is predicted by equations 1.65, the upper and lower limits of the velocity ratio to be 76 (for $Re_t<0.4$) and 8.4 (for $500 < Re_t < 2 \times 10^5$). Their results are shown in *Figure 1.19* which may thus seem to indicate the possilbe range of operating velocities for fluidized beds, i.e. from $U = U_{mf}$ where the particles are just fluidized to $U = U_t$ where the largest particles are elutriated. This is not as serious a restriction as it may appear however; most of the gas flow at velocities in excess of U_{mf} passes through the bed as bubbles which are relatively solids free, and as a result stable fluidization may be maintained at gas velocities much in excess of U_t. There will of course be entrainment and elutriation at all veclociities but properly designed cyclones can normally deal satisfactorily with any solids carried

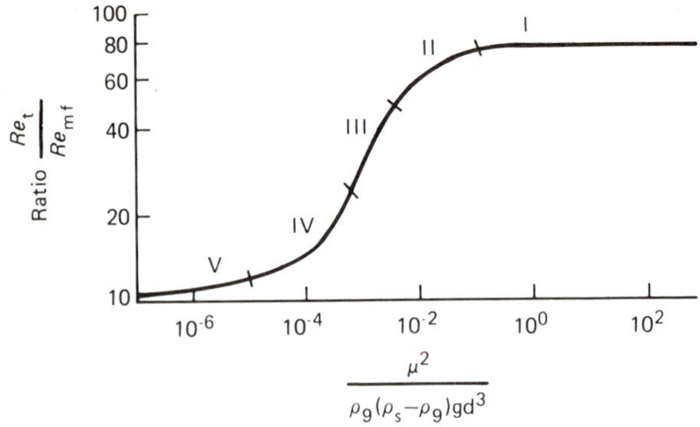

Figure 1.19 Ranges of fluidization velocity (From Pinchbeck and Popper[89])

over, and return them to the bed thereby maintaining a more-or-less constant solids inventory.

For the design of cyclones however it is necessary to know the solids elutriation rate under a given set of operating conditions and a great deal of work has been carried out in recent years on this problem. One of the most favoured approaches[90] has been to consider the rate of elutriation, $F_{i\infty}$, of solid particles within a particular size range, d_{pi}, to be directly proportional to the mass fraction of that size range, x_i, in the bed:

$$F_{i\infty} = K^* x_i \text{ (kg/m}^2\text{s)} \tag{1.67}$$

where K^* is a first-order elutriation rate constant, and further that the elutriation rate of each close size fraction from a bed of wide size distribution is unaffected by elutriation of the other size fraction, i.e.:

$$F_{\infty} = \Sigma F_{i\infty} = \Sigma K^* x_i \tag{1.68}$$

Correlations for K^* in terms of the normal operating variables have been proposed by a number of authors among whom are the following.

Table 1.1 Empirical correlations for elutriation rate constant K^*

Yagi and Aochi[91]

$$\frac{K^* g d_p^2}{\mu(U - U_t)^2} = 0.0015\, Re_t^{0.6} + 0.01\, Re_t^{1.2}$$

Wen and Hashinger[92]

$$\frac{K^*}{\rho_g(U - U_t)} = 1.52 \times 10^{-5} \left[\frac{(U - U_t)^2}{g d_p}\right]^{0.5} Re_t^{0.725} \left(\frac{\rho_s - \rho_g}{\rho_g}\right)^{1.15}$$

Merrick and Highley[93]

$$\frac{K^*}{\rho_g U} = A + 130 \exp\left[-10.4 \left(\frac{U_t}{U}\right)^{0.5} \left(\frac{U_{mf}}{U - U_{mf}}\right)^{0.25}\right]$$

Many of these empirical correlations have been criticized by Matsen[94] who has pointed out that both the Wen–Hashinger and Yagi–Aochi correlations predict an increase in entrainment with increases in both particle size and solid density and a decrease in gas viscosity. Moreover Gugnoni and Zenz[95] compared the predictions of a number of the correlations with experimental data from a bed of spherical glass beads and found the agreement in all cases to be poor. A similar point was made by George and Grace[96] who quote results from different correlations which differ by two orders of magnitude. Furthermore, the assumptions behind equation 1.68 have been brought into doubt by work reported by Geldart *et al.*[97], who found that in the absence of fines a bed of coarse particles fluidized at below the particle terminal fall velocity showed no elutriation, but that when fine particles were added continuously to the bed the coarse particles were forced into the upper reaches of the column and out into the cyclones; this was attributed to particle–particle interactions in the freeboard rather than to changes within the bed itself.

It may be concluded that elutriation correlations should be treated with the utmost caution particularly if applied outside the range for which they were determined and that a better approach to the problem would be to seek a mechanistic model on which to base predictions. A number of authors have attempted this.

A fundamental assumption in any mechanistic model of entrainment or elutriation is that concerned with the origin of the freeboard particles and, although there has been some debate as to whether they originate from the nose or the wake of rising bubbles[98], it now seems to be generally accepted that they come predominantly from the wake. Experimental evidence for this was provided by George and Grace[96] who injected single bubbles into an incipiently fluidized bed of sand, the surface of which was covered with a thin layer of coke. Entrained particles were captured in a mechanical collecting device and analysed. The fraction of coke particles collected was always a very small porportion of the whole mass of material thrown into the freeboard (less than three per cent by mass) indicating that the entrained solids originated in the deeper regions of the bed rather than at the surface and that they were probably transported up through the bed in the bubble wakes. Similar conclusions were reached by Kehoe[99] from a study of entrainment from bursting slugs.

George and Grace[96] proposed a semi-empirical model based on entrainment measurements in a 0.457 m diameter column with a number of different powders. Thus for a fluid cracking catalyst they found the volume of ejected particles (collected 0.045 m above the bed surface) normalized with respect to the volume of the bursting bubble of diameter D_b (cm) to be:

$$\xi = 0.0258 D_b - 0.014 \qquad (1.69)$$

for $3.8 \leqslant D_b \leqslant 9.0$

Then assuming the bubble flowrate Q_b to be $Q_b = \frac{1}{2}(U - U_{mf})A$, (i.e. less than that given by the two-phase theory), and using the empirical correlations of Mori and Wen[100], Geldart[101] and Werther[102] to calculate the bubble eruption diameter, they found the mass flux of particles at the bed surface to be:

$$F_o = \tfrac{1}{2}\rho_s(1 - \varepsilon_{mf})\,\xi\,(U - U_{mf}) \tag{1.70}$$

where ξ was given by equation 1.69. Entrainment above the transport disengaging height was given by:

$$F_\infty = F_o x_L \tag{1.71}$$

where x_L is the mass fraction of bed particles with terminal velocities less than the fluidizing velocity U. The prediction of TDH was based on an extension of the jet dissipation theory of Zenz and Weil[86] and led to the conclusion that:

$$\text{TDH} = 18.2\,D_b \tag{1.72}$$

The authors claim good agreement between prediction and experiment although the number and range of systems studied was somewhat limited.

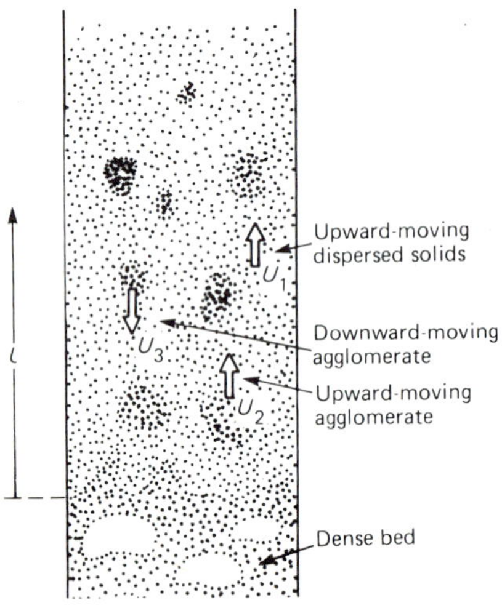

Figure 1.20 Model to account for elutriation and entrainment from fluidized beds (From Kunii and Levenspiel[31])

Kunii and Levenspiel[31] also proposed an entrainment model based on the following three postulates (*Figure 1.20*):

(i) the freeboard consists of three distinct phases, (a) gas and fully dispersed solid in pneumatic transport at velocity U_1, (b) agglomerates projected upwards with velocity U_2, (c) agglomerates descending with velocity U_3

(ii) the rate at which agglomerates are dissipated to form the dispersed phase is proportional to their concentration at any height

(iii) the rate at which upward moving agglomerates reverse direction is proportional to their concentration at the height at which reversal occurs.

Then by carrying out a mass balance on each phase and making certain simplifying assumptions, the net upward flow of solids F is found to be:

$$F = F_o e^{-aH} \tag{1.73}$$

where F_o is the flowrate of solids projected from the bed surface, H is the freeboard height and a is given by:

$$a = \frac{K_e^*}{U_2} \tag{1.74}$$

$K_e^*(s^{-1})$ being the rate constant for the transfer of solids from the upward-moving to the downard-moving phases. Although the approach seems promising it does rely on the use of empirical correlations such as the Yagi–Aochi and the Wen–Hashinger to provide numerical values of the parameters such as K_e^*.

A mechanistic model free from some of these objections has been proposed by Wen and Chen[103]. It is based on the work of George and Grace[96] and of Large *et al.*[104] and gives the solids entrainment rate at the bed surface, F_o, in terms of the diameter of erupting bubbles, D_b, and the excess gas velocity, $U - U_{mf}$:

$$\frac{F_o}{AD_b} = 3.07 \times 10^{-9} \frac{\rho_g^{3.5} g^{0.5}}{\mu^{2.5}} (U - U_{mf})^{2.5} \text{ kg/m}^5\text{s} \tag{1.75}$$

The flowrate of particles of size d_{pi} is given by a Raoult's Law type of relationship:

$$F_i = F_o x_i \tag{1.76}$$

and the variation of entrainment with freeboard height, h, by an expression similar to that of Large *et al.*[104]:

$$F = F_\infty + (F_o - F_\infty) \exp(-ah) \tag{1.77}$$

where a is an empirical constant with an average value of $4.0\,\text{m}^{-1}$. The fines elutriation rate, F_∞, is given by equation 1.67 and K^* by:

$$K^* = \rho_s(1 - \varepsilon_i)\, U_{si} \tag{1.78}$$

Here ε_i is the freeboard voidage for elutriation of single sized particles, and

U_{si} (assumed equal to $U - U_t$) is the velocity of solids of size d_{pi}. The value of ε_i is obtained from a force balance on the wall of the vessel:

$$\varepsilon_i = \left[1 + \frac{\lambda(U - U_{mf})^2}{2gD}\right]^{-0.213} \qquad (1.79)$$

where λ, the coefficient of friction between wall and particles, is found from:

$$\frac{\lambda\rho_s}{d_p^2}\left(\frac{\mu}{\rho_g}\right)^{2.5} = \begin{cases} 5.17\,Re_p^{-1.5}\,D^2 \text{ for } Re_p \leqslant Re_{pc} \\ 12.3\,Re_p^{-2.5}\,D \text{ for } Re_p \geqslant Re_{pc} \end{cases} \qquad (1.80)$$

$$Re_{pc} = 2.38/D$$

$$Re_p = \frac{\rho_g(U - U_t)d_p}{\mu}$$

The Wen–Chen correlation compares very well with a great deal of published data and has the advantage over many such predictive relationships in making specific allowance for the vessel diameter D.

1.10 Particle movement

1.10.1 Mixing

A major feature of gas fluidized beds of particles of uniform density is the high degree of solids mixing that occurs in the freely bubbling regime. Early theories of the mechanism of mixing assumed it to be the result of particle diffusion, but Rowe *et al.*[105] showed subsequently that in beds of particles of diameter greater than about 100 μm bubbles are the main agents causing particles to move through large distances, and identified two mechanisms of their action. Firstly a drift mechanism whereby a bubble draws up behind it a finger of underlying material rather in the manner of the displacement of an inviscid fluid by a rising sphere[106]. Secondly a wake transport mechanism where solids in the bubble wake are carried to the surface and replaced by solids moving downwards in the emulsion phase. *Figure 1.21* shows a section through a three-dimensional bed of two differently coloured materials of the same size and density after the passage of a single bubble; coloured particles originally occupied the lower half of the bed and white particles the upper half. The drift profile and surface deposit from the bubble wake are clearly visible; after the passage of fifty such bubbles the coloured particles were completely mixed with the rest. Rowe[107], from X-ray observations of single bubbles rising in three-dimensional beds of various powders[34], estimated the drift volume and the wake volume to be 0.35 and 0.25 times the bubble volume respectively (*Figure 1.22*). Each bubble therefore displaces upwards a

Figure 1.21 Central section through a cylindrical bed, showing the displacement caused by a single bubble (From Rowe[69])

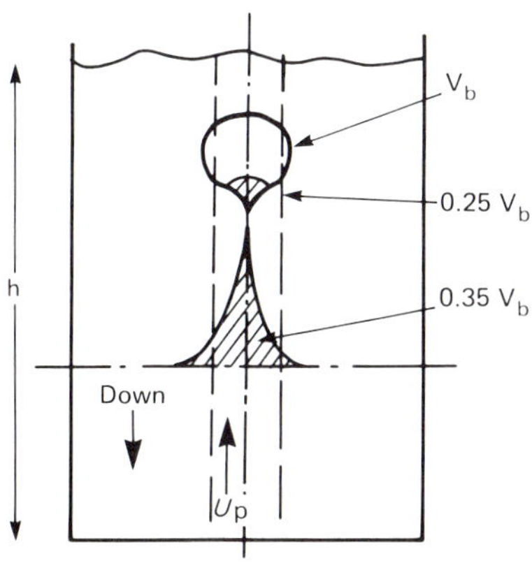

Figure 1.22 Model of particle movement (From Rowe[107])

volume of particles of about 0.6 V_b a conclusion largely in agreement with the results of Baeyens and Geldart[108] from studies with a two-dimensional bed. From continuity and applying the two-phase theory, Rowe[107] estimated the particle circulation time, t_p, to be:

$$t_p = \frac{H_{mf}}{0.6(U - U_{mf})\left[1 - \left(\frac{U - U_{mf}}{U_b}\right)\right]} \tag{1.81}$$

an equation useful for obtaining an order of magnitude estimate of whether a fluid bed with even gas distribution may be assumed to be well mixed.

Marsheck and Gomezplata[109] studied particle flow patterns in a freely bubbling 0.24 m diameter bed of cracking catalyst and showed that particles flow upwards in areas where bubbles rise, and downwards in the surrounding regions. Gabor[110] reached the same conclusions on theoretical grounds and also showed that, apart from the spread of wake solids on the bed surface following the bursting of the host bubble, little lateral mixing of particles was likely to occur. Now in small-scale units, particularly ones containing relatively large particles, there is a tendency for bubbles, as they coalesce, to move preferentially up the centre of the bed until, in beds of sufficiently high aspect ratio, they form into axisymmetric slugs. In these cases therefore solids tend to move upward in the middle of the bed and downward at the walls. Whether or not solids are well mixed is therefore strongly dependent on whether the bubble flow is distributed evenly over the cross-sectional area of the bed. In laboratory-scale units fitted with porous plate distributors and operated outside the slugging region, gas distribution is normally very uniform and consequently good particle mixing is observed. On the industrial scale however where gas may be introduced into a bed through regularly distributed but discreet entry points (nozzles, tuyeres, etc.) there is always the possibility that bubbles will rise along preferred paths and so cause regular solid circulation patterns to be established, thereby inhibiting complete mixing.

Evidence for this latter effect has come from the extensive work of Whitehead, Potter *et al.* with 0.61 m and 1.22 m square beds fitted with a variety and varying number of gas inlet tuyeres[111–114]. One series of reported tests was carried out with the larger bed fitted with 36 tuyeres set in a square array at 0.20 m centres[113]. The bed material was silica sand (U_{mf} = 25 mm/s), of varied static bed heights and fluidized at different gas velocities. The bed was defluidized at varied time intervals after quantities of tracer particles had been added, and the slumped bed was then sectioned to reveal the positions reached by the tracers. The experiments showed that under certain conditions of free bubbling a highly ordered circulation pattern was set up with a predominant downflow in the centre of the bed and a smaller downflow near the walls; the pattern is shown in idealized form in *Figure 1.23*. The downflow velocity of the central portion

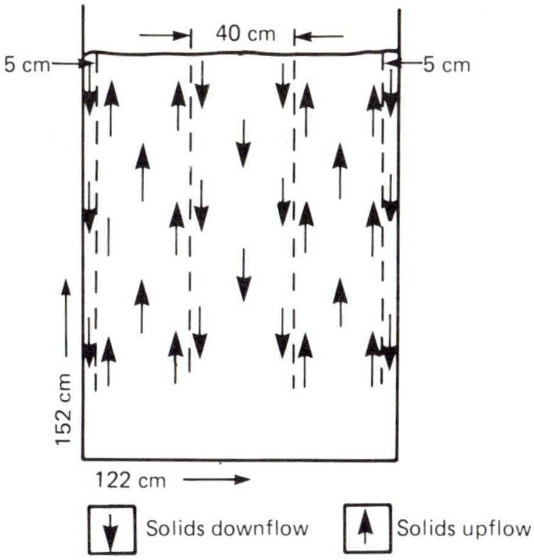

Figure 1.23 Schematic diagram of one particular type of solids circulation pattern (From Whitehead et al.[113])

of the stream was found to be of the order of 0.21 m/s in a 1.47 m deep bed fluidized at a gas velocity of some $6U_{mf}$. The rate of lateral solids mixing was found to be influenced by bed height, tracer particles added to a 1.47 m deep bed having become well dispersed 15 seconds after their addition while with a 0.40 m deep bed at the same velocity complete dispersion had not been achieved after 25 seconds.

A similar but rather more extreme example of this lack of good lateral mixing was reported by Singh *et al.*[115] who were interested in burning gas in a rectangular bed 0.91 m × 0.61 m and 0.50 m deep. They found that solid tracers could be completely mixed vertically in 75 seconds at a fluidizing velocity of $3U_{mf}$, but that complete mixing over the entire length of the bed required some 1.5 hours. Whitehead and Potter also considered the effect of particle downflow on the mixing of emulsion-phase gas[116,117]. Their experiments showed that under some conditions gas backmixing can begin to develop at velocities as low as $2.6U_{mf}$, a figure in general agreement with the 'counter current backmixing' model of chemical reaction developed earlier by Potter *et al.* (*see* Chapter 2, Section 2.3.5).

Based on the experiments of Rowe *et al.*[105], Werther has developed an empirical correlation for the average convective solids mass flow per unit bed area, F(kg/m^{2}s), of Geldart Group B powders in large scale beds[118]. Thus in SI units:

$$F = 0.67\,(1 - \varepsilon_{mf})\rho_s \left(\frac{1}{\phi_b} - 1\right)(U - U_{mf}) \tag{1.82}$$

where the dimensionless bubble shape factor ϕ_b is given by:

$$\phi_b = \{1 - 0.3\exp[-8(U - U_{mf})]\}\exp(-\sigma h) \tag{1.83}$$

$$\text{where } \sigma = 7.2(U - U_{mf})\exp[-4.1(U - U_{mf})] \tag{1.84}$$

and h is the height above the distributor. Very good agreement was found between the predictions of equation 1.82 and the measurements of Schmalfeld[119] on solid circulation in a pilot-scale (0.8 m diameter) semi-cylindrical bed.

The motion of solid particles in a fluidized bed can also be influenced by arranging for an uneven distribution of gas flow through the distributor plate. Thus Merry and Davidson[120,121] induced gross circulation of particles in both two- and three-dimensional beds by supplying more gas to the outer regions of the distributor than to the central area (*Figure 1.24*). In

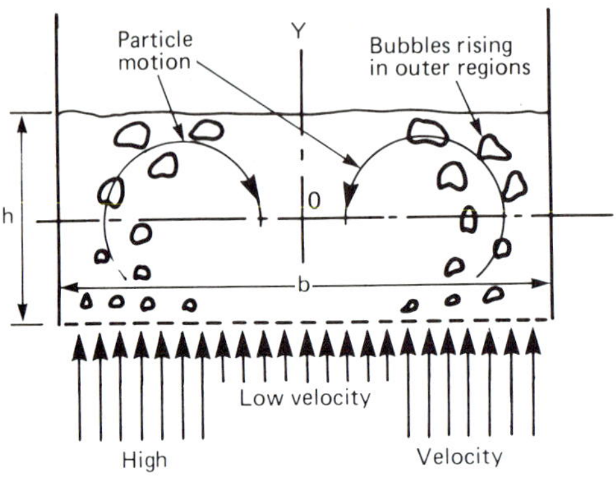

Figure 1.24 Pair of two-dimensional cells showing mechanism of 'Gulf Stream' circulation (From Merry and Davidson[121])

this way an excess of particles was moved upwards in bubble wakes in the outer regions while in the centre of the bed, due to the relative scarcity of bubbles in that region, particles were moved predominantly downwards. With the arrangement shown particles were induced to circulate continuously in a steady pattern in two circulation cells–the so-called 'Gulf Stream' circulation. This type of motion is favoured particularly in relatively shallow beds. If the bed is too deep the bubbles tend to become more evenly spread out across the bed in the surface region. On the other hand a bed can be too shallow for the circulation to become stabilized. Merry and Davidson[120] found the optimum ratio of bed height, h, to width, b, in their essentially 2-D beds, to be $h/b = 0.41$. In experiments with a 0.61 m square bed[121], again with an excess of gas flow through the outer regions of the distributor plate, it was found that Gulf Stream circulation could be set up for a range of bed depths. With a 0.15 m deep bed ($h/b = 0.25$) little circulation was observed and the downward velocity of particles in the centre of the bed was effectively zero. As bed depth was increased however, circulation was induced until at a bed depth of 0.35 m ($h/b = 0.61$) the rate of circulation began to decrease again. Although the authors established some quantitative relationships for these systems, for example between circulation velocity and bed dimensions:

$$U_p = \frac{2bh^2 \zeta_o}{\pi(b^2 + 4h^2)} \qquad (1.85)$$

where ζ_0 is the vorticity at the centre of circulation, the applicability of the equations to large-scale units is at present uncertain. There is no doubt however about the importance of the principle of this mode of solids mixing for large industrial units where solids need to be supplied to a fluid bed from a number of discrete points and mixed as quickly and efficiently as possible with the bulk of the bed. An example of a system where this is a prime requirement is a fluid bed combustor for burning coal or other solids with a high volatiles content where, without some kind of forced circulation, the combustible material tends to float on the bed surface and to burn there in an uncontrollable manner. Gulf Stream circulation offers a simple and straightforward solution to the problem provided the dimensions of the circulation cells can be determined in advance (*see* Section 4.4.1).

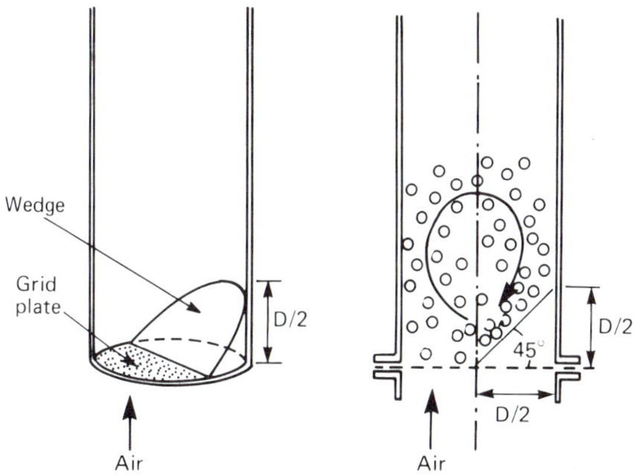

Figure 1.25 Whirling bed apparatus (From Rios et al.[75])

The induced circulation of large solid particles such as cocoa beans has been studied by Rios *et al.*[75] using a device known as a 'whirling bed', (*Figure 1.25*). In this system the perforated plate distributor is half covered by a 45° angle wedge which serves to return particles, which are effectively defluidized in the upper region of the bed, to the grid where they are again carried upwards by the entering air. The solids motion is closer to that observed in a spouted bed than in a fluidized bed (there are no bubbles visible at the wall of the vessel) but it provides a very efficient means of bringing a gas and large particles into intimate contact. It is currently under investigation for use in the food industry for the drying or blanching of edible solids.

1.10.2 Segregation

When a bed containing a mixture of particles of different size or density is fluidized it is often observed that the particles segregate, the bigger or denser sinking to the bottom of the bed while the smaller or lighter remain in the upper regions. Rowe *et al.*[122] have used the term 'flotsam' to describe the material remaining near the bed surface and 'jetsam'[a] to describe that which sinks to the bottom. From observations with a two-dimensional bed these workers concluded that density difference was the major factor in promoting segregation and that while different sized particles of the same density also segregate, the larger being the jetsam, the effect of size difference is slight, particles differing in diameter by as much as an order of magnitude mixing uniformly at moderate gas flowrates.

For spherical particles of different density Rowe *et al.*[122] were able to distinguish three basic mechanisms by which mixing and secregation occurred. The most important mechanism was one in which particles from the bottom of the bed, both flotsam and jetsam, were lifted to the top in the wake of a rising bubble; this was the sole mode of transport of small, light particles towards the surface. The movement of jetsam to the bottom of the bed occurred by its falling through the free space inside bubbles. This latter effect was particularly important at gas velocities sufficient to fluidize the lighter particles but not the heavier. A third mechanism, also leading to segregation, was observed with dense particles that were much smaller than their lighter, less dense partners; in these cases it was possible for the jetsam to sink by percolating slowly downwards through the emulsion phase but the tracks followed were generally confined to the path of a recent bubble.

The segregation patterns observed by Nienow *et al.*[123] are shown in *Figure 1.26 (c)&(d)*, which also illustrates the definition of a mixing index, $M = x/\bar{x}$, that varies from zero for complete segregation to 1 for complete mixing (*Figure 1.26 (a)&(b)*. Here x is the mass fraction of jetsam in the upper uniform part of the bed, while $\bar{x}$ is the overall mass fraction of jetsam in the bed. From these observations it is apparent that even at high gas velocities some of the denser material remains uniformly distributed throughout the upper layer, the proportion increasing with gas flowrate. Measurement of the mixing index M over a range of gas velocities showed it to vary in a logistic manner (*Figure 1.27*):

$$M = \frac{1}{(1 + e^{-Z})} \tag{1.86}$$

where Z is a velocity ratio defined by:

$$Z = \left(\frac{U - U_{TO}}{U - U_F}\right) \exp\left(\frac{U}{U_{TO}}\right) \tag{1.87}$$

[a]Strictly speaking this is a misnomer. Jetsam is defined in the *Concise Oxford Dictionary*, 5th Edn (1964), as 'goods thrown overboard from a ship to lighten it'. A more appropriate word would be 'lagen', defined (op. cit.) as 'goods or wreckage lying on the bed of the sea'.

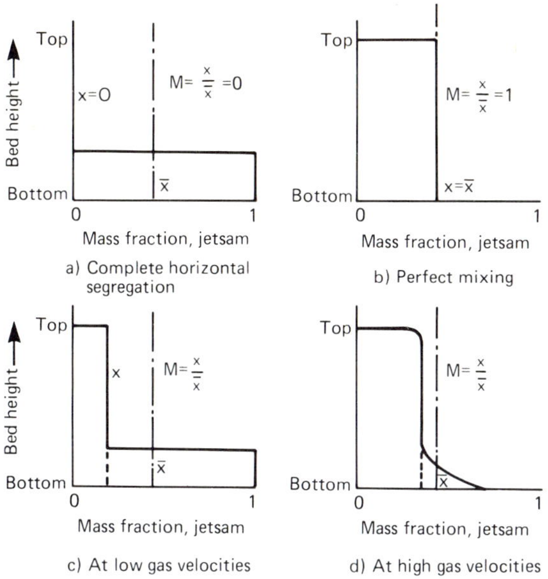

Figure 1.26 Segregation patterns and mixing index for close-sieved binary mixtures (From Nienow et al.[123])

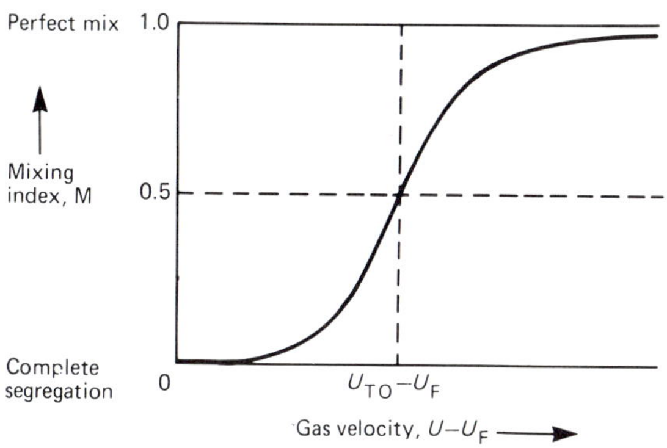

Figure 1.27 Variation of mixing index with excess gas velocity (From Nienow et al.[123])

Here U_F is the minimum fluidization velocity of the lighter component and U_{TO} is the 'take-over' velocity at which $M = 0.5$. Below U_{TO} segregation predominates while above this value mixing takes over. An empirical expression was found relating U_{TO} to the operating parameters[124] thus:

$$\frac{U_{TO}}{U_F} = \left(\frac{U_P}{U_F}\right)^{1.2} + 0.9(\rho_R - 1)^{1.1}(d_{ER})^{0.7} - 2.2(\bar{x})^{0.5}(H^*)^{1.4} \quad (1.88)$$

where:

U_P = minimum fluidization velocity of the denser material

$\rho_R = \dfrac{\rho_H}{\rho_L}$

ρ_H = density of the heavier material

ρ_L = density of the lighter material

d_{ER} = shape-corrected diameter ratio $= \dfrac{\phi_H d_H}{\phi_L d_L}$

ϕ = sphericity

H^* $= 1 - \exp(-H/D)$

H = bed height

D = bed diameter

A theoretical model of particle segregation was developed by Gibilaro and Rowe[125]. The model is based on four physical mechanisms: particle circulation, wake-bulk phase interchange, axial dispersion and particle segregation. Comparison with experiment shows that interphase exchange is generally small and can be neglected, and that furthermore axial mixing is only important at high gas velocities or in weakly segregating systems. Many of the predictions of this model have yet to be verified experimentally.

The foregoing remarks apply exclusively to binary systems, and so far little work has been carried out on predicting the behaviour of more complex mixtures.

Another factor influencing the mixing and segregation patterns of solids is particle shape. Again, little has been reported in the literature on this although Nienow and Cheesman[126] reported a preliminary study which indicated little influence of shape except in cases where large, flat particles of relatively low density remained on or near the distributor and were orientated parallel to the gas flow.

1.11 Bed viscosity

Owing to the many similarities between fluidized beds and liquids, a number of workers have found it useful to attribute the property of viscosity to gas–solid systems. This is an attractive idea and one that can be of considerable utility when examining fluid bed hydrodynamics and stability, although strictly speaking, the concept of viscosity resulting from gas–particle and/or particle–particle shearing forces is only applicable to the emulsion phase of the bed since the rising bubbles are essentially devoid of particles. A problem arises however when numerical values of viscosity are required, as it is very doubtful whether any conventional viscometer can be used in a fluidized bed without interfering in a fundamental way with the bed's gas–solid contacting pattern and so altering the very property it is designed to measure. It has therefore been necessary to use indirect methods of estimation that do not involve any interference with the bed and although a number of such approaches have been attempted the method proposed by Grace[127] has been the most successful.

Grace's method consists of estimating viscosity from the shape of bubbles in fluidized beds. Thus observations of rising gas bubbles in liquids show them to be spherical, ellipsoidal or spherical-cap shaped[128–130]. The

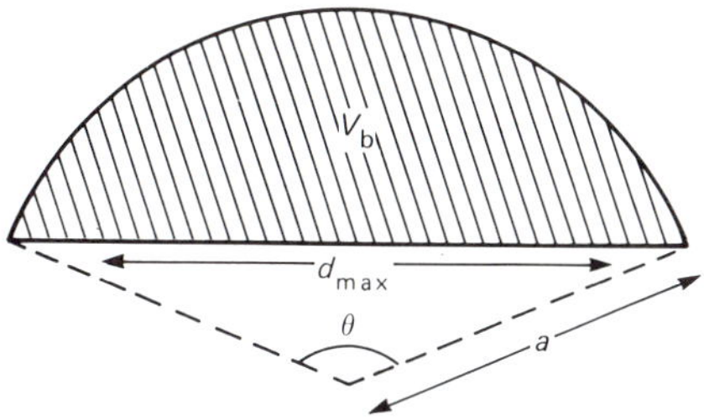

Figure 1.28 Schematic diagram of spherical-cap bubble

latter (*Figure 1.28*) are typically formed in liquids of low viscosity and have an included angle, θ, of about 100°[41]. As the liquid viscosity increases however the angle θ also increases and the bubbles tend to become more spherical in shape. This led Grace to propose a formal link between viscosity and the included angle of a gas bubble, and on the basis of the empirical observations of a number of authors[131–135] he obtained the relationship:

$$Re_b = 23 \exp(-0.004\theta) \quad (1.89)$$

where Re_b is a bubble Reynolds number given by:

$$Re_b = \frac{d_b U_b \rho_s (1 - \varepsilon_{mf})}{\mu_A} \quad (1.90)$$

Here d_b is the diameter of the sphere of volume V_b, U_b is the bubble rise velocity (calculated from the Davies–Taylor relationship $U_b = \frac{2}{3} g^{1/2} a^{1/2}$), ρ_s is the particle density and μ_A is the apparent viscosity of the emulsion phase which is assumed to have a voidage ε_{mf} equivalent to that at minimum fluidization. The experimental results on which the correlation was based were chosen such that the bubbles were truly spherical-capped, surface tension forces were small compared with inertia and bouyancy forces, and wall effects were negligible (maximum bubble dimension, d_{max}, less than one fifth of the bed diameter). These conditions were satisfied by X-ray observations of gas bubbles in three-dimensional fluid beds[34,136] from which included angles of 200–260° were measured. Based on these values bed viscosities were calculated using the empirical equation 1.90, and the results so obtained are given in *Table 1.2*. These values, ranging from 0.4 to 1.3 Ns/m^2 are generally very much higher than can be accounted for on the basis of gas–particle shearing forces alone. Thus

Table 1.2 Apparent viscosities determined by the method of Grace[127]

Material	*Size* (μm)	*Shape*	*Size range*	*Apparent viscosity* (Ns/m^2)
Ballotini	550	spherical	narrow	0.95
	220			0.85
	170			0.75
	140			0.80
	120			0.85
	82			0.90
	60			0.70
Silver sand	500	angular	close	1.2
	460			1.0
	330			1.3
	230			0.7
	140			0.9
	82			0.85
	72			0.8
Synclyst	52	rounded, irregular	wide	0.4
Magnesite	240	angular	wide	0.9

according to the treatment of Frankel and Acrivos[137], the maximum effective viscosity of a concentrated suspension of solid spheres assuming no particle–particle interactions should be only about 100 times the viscosity of the pure fluid, which for air at ambient conditions ($\mu = 1.7 \times 10^{-5} Ns/m^2$) leads to a value some three orders of magnitude lower than those determined by Grace's method. The conclusion to be drawn from this is that particle interactions must be an important factor in determining the viscosity of fluidized beds.

1.12 Fluidization at elevated pressure

For many industrial processes that use fluidized beds, operation at atmospheric pressure is quite adequate. There are a small number of applications however where high pressure operation can confer considerable advantages and since at least two of these, coal combustion and gasification, are in the forefront of fluidization technology there is currently a great deal of interest in pressurized fluidization. Although coal processing is dealt with in detail in Chapter 4, it is worthwhile here to summarize the benefits of operating a combustor at high pressure. Perhaps the major advantage is the potential reduction in plant size, and hence cost, that can be achieved for a given heat output, since the dimensions of a fluid bed combustion unit are determined partly by the quantity of air that must be supplied to burn the fuel. Pressurized operation thus permits a proportional reduction in the volumetric air supply which in turn leads to a corresponding reduction in the required bed area.

A second advantage is the increase in overall efficiency that is made possible by expanding the high pressure off-gases from the bed through a gas turbine. A number of operating cycles based on this principle have been proposed although the problem of removing enough of the entrained bed particles to make the off-gas suitable for supply to a turbine has not yet been solved. Using currently available turbine technology the fluid bed would need to be operated somewhere in the region of 20 bar pressure. Other advantages of pressurized combustion are enhanced bed-to-surface heat transfer, increased sulphur retention on limestone and decreased emission of nitrogen oxides.

In the case of pressurized coal gasification the advantages lie not only in the reduction in equipment size and cost for a given throughput but also in the thermodynamic favourability of the reaction forming methane from carbon and hydrogen. For this latter application pressures of the order of 70–100 bar are being considered.

There is thus considerable industrial importance attached to fluidization at elevated pressures, but despite a recent marked increase in research activity there is still much that is not understood about this mode of operation. In many reported studies, reference is made to the improved 'quality' or 'smoothness' of fluidization under pressure[56,138–140] and this has led to comparisons being made between the behaviour of pressurized gas–solid systems and liquid–solid beds. Thus Godard and Richardson[141] fluidized diakon powder ($\rho_s = 1.18 \times 10^3$ kg/m^3; $\bar{d}_p = 125$ μm) and phenolic resin particles ($\rho_s = 0.24 \times 10^3$ kg/m^3; $\bar{d}_p = 121$ μm) in a 0.10 m diameter Perspex tube at pressures from 1 to 14 bar and found an increased range of bed expansion between the points of minimum fluidization and minimum bubbling. The velocity (U)–voidage (ε) relationship in this range was identical to the Richardson–Zaki equation typical for liquid–solid beds:

$$\frac{U}{U_i} = \varepsilon^n \tag{1.29}$$

Here U_i was about equal to the terminal fall velocity, U_t, for the lighter particles but approached $2U_t$ for the diakon. Furthermore the value of the exponent n was somewhat higher than the value found for uniform spheres in a liquid. In general, n was found to decrease and ε to increase with increasing pressure, although the change in voidage was more marked between 1 and 8 bar than at the higher pressures.

The improved qualtiy of gas–solid fluidization under pressure has also been attributed to the frequently-made observation that for a given flow velocity the gas bubbles present are much smaller than those found at atmospheric pressure. This has led to a reopening of the debate about the stability of gas bubbles in fluid beds and the theories mentioned earlier have been re-examined to see what effect pressure would be expected to have on their predictions.

Guedes de Carvalho and Harrison[10] modified the earlier theory of Harrison *et al.*[63] which assumes that gas circulates within a bubble at a velocity, U_c, which approximates to the bubble rise velocity, U_b, and that when U_b exceeds the terminal fall velocity of the bed particles, wake solids are drawn up into the bubble causing it to break up into a number of smaller bubbles each with a lower rise velocity. Now if gas is passed through a horizontal tube the floor of which is covered with solid particles, at a certain gas velocity known as the 'saltation velocity', U_s, the particles will be picked up by the gas and transported along with it. The process has been studied by Zenz[142] who gives plots showing how U_s may be calculated from the physical properties of gas and solids, and from these it is clear that U_s decreases strongly with gas density and so with pressure. The modification suggested by Guedes de Carvalho and Harrison was that bubble break-up occurs when U_b becomes equal to U_s, i.e. when the velocity of gas circulation within the bubble is sufficient to lift particles from the wake into the void above it. They pointed out that since U_s decreases with pressure the maximum stable size of bubble should also decrease. Hence from plots of U_t and U_s against pressure they concluded that a slug of air rising at 0.245 m/s in a 0.05 m diameter bed of synclyst catalyst powder (ρ_s = 890 kg/m^3; $\bar{d}_p$ = 48 μm) should cease to be stable at a pressure of 15 bar, whereas in a bed of glass ballotini (ρ_s = 2920 kg/m^3; $\bar{d}_p$ = 68 μm) it should remain stable at up to 30 bar. They fluidized beds of these materials at pressures of up to 22 bar and found that for a range of gas velocities slugs were formed at all pressures in the bed of ballotini, but that over the same velocity range the bed of synclyst ceased to slug at 6 bar and remained in the bubbling regime at all pressures up to 22 bar. Allowing for the approximations involved in the calculations of U_s, the experimental results were concluded to be in good agreement with the theory. Further experimental work along the same lines was reported by Guedes de Carvalho *et al.*[143] and subsequently by King and Harrison[144]. In the latter study, X-ray cine photographs were taken of rising slugs in pressurized 0.1 m diameter beds of ten different powders divided between Geldart Groups A and B, and two conclusions of major importance were drawn. Firstly it was clear from an examination of the films that when the slugs broke up they did so in all cases by fingers of particles falling from the leading edge of the slug and not by becoming filled with particles from the wake. Secondly it was shown that up to 25 bar, pressure had little effect on Group B powders; the bubbles and slugs formed in them remained stable and the two-phase theory was a good approximation to the observed flow division in all cases. With beds of Group A powders however bubbles and slugs became increasingly unstable at the nose at elevated pressures and above 5 bar the two-phase theory was no longer valid as more gas than predicted flowed through the emulsion phase. The results of this work would thus seem to argue against the theory that bubble instability is caused by particles filling the void from the wake and to suggest that alternative explanations for the behaviour of fluid beds under pressure must be found. The theory of Taylor instabilities at the leading edge of a bubble developed by Upson and Pyle[65] and Clift *et al.*[66] has already been mentioned and its proposed mechanism of bubble break-up would appear to have received support from the observations of King and Harrison. Further support has come from the work reported by Subzwari *et al.*[145] who found an increase in emulsion phase voidage and a reduction in bubble flow at pressures above 5 bar in a two-dimensional bed of fluidized silica–alumina cracking catalyst. They also found that the dominant mechanism of bubble splitting irrespective of pressure was by division from the roof and that the maximum stable bubble size decreased with increasing pressure. Now the Taylor instability theory predicts that bubbles will become less stable as the kinematic viscosity of the emulsion phase decreases i.e. that smaller bubbles will result from an increase in dense phase voidage. Prediction and observation are thus in at least qualitative agreement but the reason for the dense phase expanding in the first instance is still not clear.

Rowe and MacGillivray[146] using an X-ray technique, made observations at 1 and 4 bar pressure with a fine powder ($\bar{d}_p = 58$ μm) in a 0.2 m diameter vessel and found that for the same volumetric gas flowrate the mean bubble volume was reduced by 30 per cent at the higher pressure. They also reported an increase in average bubble velocity of about 20 per cent at 4 bar.

The effect of pressure on two-dimensional beds of sand particles 0.3, 0.45 and 0.6 mm in diameter was reported by Kawabata *et al.*[147]. They found that the minimum fluidization velocity of these coarse particles decreased with pressure up to 8 bar and that the reduction was explained readily on the basis of the Egun equation. The second term of this equation (equation 1.11) corresponds to the turbulent flow regime and it becomes appreciable as pressure increases. Furthermore the equation predicts that the coarser the particles are the greater is the effect of pressure, a trend that was seen experimentally. The Japanese workers also observed a flattening of the bubbles with increasing pressure and used this effect to account for a decrease in bubble rise velocity leading to an increase in bubble hold up and hence to an increase in bed expansion. There is some evidence from X-ray studies[148] that bubble flattening also occurs in three-dimensional beds of coarse particles at 60 bar and although the details of the process have yet to be clarified it would seem to be caused by wake particles intruding into the bubble space above them. In other words the visual evidence from these two latter studies would seem to favour the idea that wake particles do tend to fill the bubble space as pressure increases and in the light of this, the question of bubble stability and its variation with pressure appears to be in need of re-examination.

1.13 Heat transfer

1.13.1 Bed-to-surface heat transfer

In many fluidized bed processes it is necessary to transfer heat between the bed material and a surface. The latter can be an immersed cooling or heating coil through which heat transfer fluid is circulated or it can be the wall of the bed itself. In general this heat transfer process is very efficient and even when carrying out the most exothermic of chemical reactions in a fluid bed it is not normally difficult to achieve fully isothermal operation. These excellent heat transfer characteristics have been appreciated since the earliest applications of the fluid bed technique and they continue to be one of its major advantages for large scale operation in the process industries. The variation with gas velocity of the bed-to-surface heat transfer coefficient under conditions where radiation transfer may be neglected, is shown in *Figure 1.29.* It is apparent that the curve may be

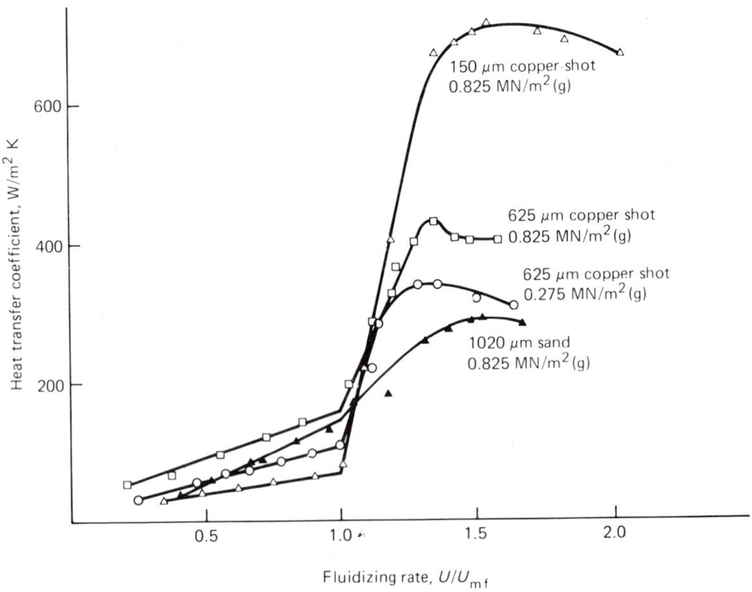

Figure 1.29 Variation in wall-to-bed heat transfer coefficient as a function of fluidizing gas flowrate under conditions where radiant heat transfer may be neglected (From Botterill[149])

divided into three regions, (i) below U_{mf} where the bed is in the packed state and the heat transfer coefficient is low, (ii) between U_{mf} and an optimum velocity where the coefficient increases sharply to a maximum value, (iii) above the optimum velocity where a gradual decrease sets in. It is generally believed that the sharp increase in heat transfer coefficient above U_{mf} is caused by bubble-induced particle motion at the transfer surface while the decrease above the optimum velocity is the result of restricted particle–surface contact under conditions of high bubble flow. It is also apparent from *Figure 1.29* that heat transfer coefficients increase with increasing particle thermal conductivity and with decreasing particle size although the exact form of the interdependence of these and the other operational parameters is clearly quite complex.

Because of its importance, an enormous amount of research has been carried out into the subject; a review by Botterill[149] contains some three hundred references to work in the open literature up to 1975. Much of the early work in this area was directed towards obtaining empirical correlations between heat transfer coefficients and the physical properties of the gas–solid system based on a steady-state model which assumed that heat is transferred through a laminar gas film whose thickness is a function of the velocity of the solid particles as they 'scour' the surface; (*see* for example Levenspiel and Walton[150], Wen and Leva[151], Wender and Cooper[152] and Vreedenberg[153]). Following the studies of Mickley and Trilling[154] and Mickley and Fairbanks[155] however it became clear that in the absence of any radiant component (below 600°C) bed-to-surface heat transfer is better represented as an unsteady-state process in which mobile 'packets' of emulsion phase material carry heat to or from the surface, residing there for a short period of time before moving back into the bulk of the bed and being replaced by fresh material. The rate of heat transfer was predicted to be a maximum at the instant of contact but to decrease rapidly as the residence time of the packet at the surface increased and the local temperature gradient was reduced. It was accepted that the surface renewal was a direct result of the passage through the bed of gas bubbles and their attendant wakes. Mickley *et al.*[156] proposed the following expression for the bed-to-surface heat transfer coefficient, h_{bs}

$$h_{bs} = 2\left[\frac{k_e \rho (1 - \varepsilon) C_e}{\pi \bar{t}}\right]^{1/2} \tag{1.91}$$

where k_e is the thermal conductivity, ε the voidage and C_e the average heat capacity of the emulsion phase, ρ is the particle density and $\bar{t}$ is the mean residence time of the packets at the surface. The quantity of heat transferred when the surface was immersed in a gas bubble was considered to be negligibly small. The model was modified by Baskakov[157] who introduced the idea of a contact resistance at the wall caused by the higher void fraction in the surface region compared with the bulk of the emulsion phase. Gelperin and Einstein[158] developed Baskakov's theory and proposed that the contact resistance, R_c, should depend on the effective conductivity of the emulsion in the surface zone, k_{ew}, and on the thickness of the zone, δ:

$$R_c = \frac{\delta}{k_{ew}} \tag{1.92}$$

The value of δ was assumed to be of the order of the particle diameter, d_p.

An alternative approach closely linked to theories of heat transfer in packed beds was proposed by a number of authors who suggested that the packets of emulsion at the transfer surface and the bulk emulsion phase itself could not be regarded as having uniform thermal properties but that the two should be considered separately. Botterill and Williams[159] were among the first to explore this concept and they proposed a heat transfer mechanism that involved simultaneous conduction from a flat surface to both interstitial gas and a single layer of particles through their points of contact. They solved the unsteady state conduction equations for this model and obtained isotherms for varying contact times and for different gas and solid properties. They showed that owing to their very high heat capacity relative to that of the surrounding gas, solid particles act as a local heat sink and are able to ensure the rapid establishment of thermal equilibrium between surface and bed. Experimental test of the theory[160] was carried out using a system fitted with a rotating paddle which scoured the inside wall of the fluid bed and enabled particle residence times at the wall to be varied. The tests showed however that the predicted heat transfer coefficients were up to an order of magnitude too high, a result that was attributed to the presence of a gas film some 10 μm thick separating the particles from the surface. Similar results were obtained by Couderc *et al.*[161] who reconciled theory and experiment by assuming a gas layer of maximum thickness $d_p/5$ at the wall. Gabor[162] proposed a model on the same lines and explored the consequences of assuming heat to be transferred through a string of particles normal to the wall surface. He concluded that a contact resistance was not necessary to obtain agreement between theory and experiment.

In recent years the two approaches have moved closer together and it now sees to be generally accepted that heat transfer to immersed surfaces may be considered to a first approximation to be made up of three additive components[149,163]:

(i) the particle convective component accounting for the heat transferred to aggregates of particles whose motion is a direct result of bubble flow
(ii) the gas convective component accounting for transfer to emulsion phase gas
(iii) the radiant component that becomes important only at temperatures in excess of 600°C.

A model incorporating the first two of these components has recently been developed by Xavier and Davidson[163] and applied to a low-temperature slugging bed for which the hydrodynamics are well defined. They based an evaluation of the particle convective transfer component, h_{pc}, on the Mickley–Fairbanks model incorporating the Baskakov modification to take account of gas film resistance at the wall. The particle packet contact time, $\bar{t}$, for slug flow past a surface of length L was given by:

$$\bar{t} = \frac{L}{U - U_{mf}} \tag{1.93}$$

and hence the transfer coefficient was found to be:

$$h_{pc} = 2\left[\frac{k_{mf}\,\rho_{mf}\,C_{mf}(U - U_{mf})}{\pi L}\right]^{1/2} \frac{0.35(gD)^{1/2}}{U - U_{mf} + 0.35(gD)^{1/2}} \tag{1.94}$$

where k_{mf}, ρ_{mf} and C_{mf} refer to the conductivity, density and heat capacity of the emulsion phase at U_{mf}, and D is the column diameter. The equation was found to hold for small particles ($\bar{d}_p$<100 μm) and long contact times (L>100 mm) with the main parameters calculated as follows:

(i) k_{mf} was obtained from the packed bed relationships of Yagi and Kunii[164] and Yagi *et al.*[165]:

$$k_{mf} = k^0{}_e + 0.1\,\rho_g C_g\, d_p\, U_{mf} \tag{1.95}$$

where $k^0{}_e$ was found from Kunii and Smith[166] or from Gelperin and Einstein[158]:

(ii) ρ_{mf} was equated to $\rho_s\,(1 - \varepsilon_{mf})$
(iii) C_{mf} was taken as equal to the heat capacity of the solid particles.

The Baskakov contact resistance, R_{co}, was used to evaluate a film transfer coefficient h_{film} which becomes important as particle size increases:

$$\frac{1}{R_{co}} = h_{film} = \frac{k_{gt}}{\delta} \tag{1.96}$$

where the film thickness δ is assumed to be in the range $d_p/10<\delta<d_p/6$ and k_{gt} is the thermal conductivity of the fluidizing gas. The general particle convective heat transfer coefficient was then shown to be:

$$h_{pc} = h_{pc,max}\left[\frac{(\pi x)^{1/2}}{y(1 + x)}\ (\text{erfc}\ y.\exp\ y^2 - 1) + \frac{2x^{1/2}}{(1 + x)}\right] \tag{1.97}$$

where $y = \dfrac{h_{film}}{h_{pc,max}(\pi x)^{1/2}}$ and $x = \dfrac{U - U_{mf}}{0.35(gD)^{1/2}}$

Xavier and Davidson obtained the gas convective component of heat transfer, h_{gc}, from the model of Gabor[167] for a flat surface:

$$h_{gc} = \left(\frac{4\,k_{mf}\,\rho_g\,C_g\,U_{mf}}{\pi L}\right)^{1/2} \tag{1.98}$$

and assumed the overall transfer coefficient to be:

$$h_c = h_{pc} + h_{gc} \tag{1.99}$$

From these equations, derived for slugging beds, the authors proposed tentative correlations for bubbling systems in terms of the dimensionless groups $h_c/h_{pc,max}$ and $h_{pc,max}/h_{film}$ for small particles, and h_c/h_{gc} and h_{gc}/h_{film} for large particles and found reasonable agreement between model predictions and experimental values. The work has recently been extended by Xavier *et al.*[168] to high-pressure beds and again reasonable agreement was found at pressure of 1–20 bar; at 25 bar however the theoretical estimates were some 20 per cent too low, a result attributed to the breakdown of well-defined bubble/slugs at this pressure. The model would thus seem to be restricted in applicability to well-behaved systems although for such systems it has the considerable virtue of being simple and founded on generally accepted bubble mechanics.

1.13.2 Particle-to-fluid heat transfer

Owing to the fact that the flow of gas around particles in a fluidized bed is essentially streamline and that the local Reynolds number is correspondingly small, particle-to-fluid heat transfer coefficients are normally quite low; Botterill[149] quotes values of 6–23 W/m^{2}K. However, because the bed particles are small and numerous, the total surface area of solid per unit volume of bed exposed to the flowing gas is very large indeed (3000–45 000 m^2/m^3)[149], with the result that the overall heat transfer rate between particles and gas is very rapid. This, added to the much larger heat capacities of solids than gases, causes a fast equilibration in temperature between an incoming gas and the bed material so that, within a few particle diameters of the point of entry of the gas, heat will have been exchanged totally and the gas will have attained the temperature of the bulk of the bed. It is for this reason that fluidized beds are, more often than not, completely isothermal, departures from the norm only arising in situations where there is maldistribution of gas leading to the formation of defluidized zones within the bed.

1.14 Gas distributors

In most instances fluidizing gas is introduced into a bed of particles through a distributor or grid that spans the cross-section of the bed at its base. In small-scale work carried out with laboratory-sized units distributors are often in the form of a flat porous plate made from sintered metal or ceramic material but on an industrial scale they are more commonly constructed from metal plates with a number of discrete holes arranged in a regular geometric pattern. In these cases the holes may be located in simple nozzles or the plate perforations may be covered with a bubble cap or tuyere that serves to prevent solid particles from leaking back into the space below (*Figure 1.30*). The space below the grid is called the 'windbox' or 'plenum chamber' and should be sufficiently large volume to ensure that a constant gas pressure is maintained within it. Some industrial designs of distributor are described in detail in the book by Kunii and Levenspiel[31].

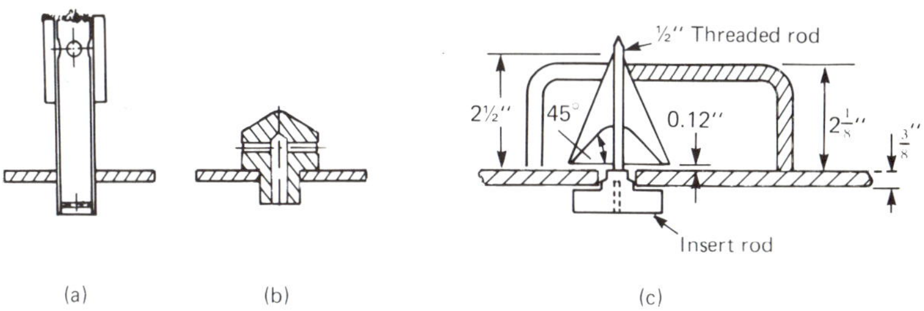

Figure 1.30 Distributor gas inlets: (a) tuyere (from Whitehead and Young[51]), (b) nozzle (from Werther[169]), (c) cap (from Petrie and Black[170])

An essential function of the distributor is to introduce the fluidizing gas evenly through each hole thereby keeping the solid particles above it in constant motion and preventing the formation of defluidized zones within the bed, and to this end it is necessary to maintain a sufficiently high pressure drop across the grid. The magnitude of this pressure drop, Δp_D, is generally considered in relation to that of the pressure drop across the bed itself, Δp_B, but actual values recommended in the literature vary widely. Thus Agarwal *et al.*[171] from experience with an industrial coal dryer, specified a minimum value for Δp_D of 10 per cent of the bed pressure drop whereas Zenz[172] recommended that it should be at least 30 per cent. Zuiderweg[173] proposed that for a perforated plate or nozzle distributor the pressure drop should be related to both minimum fluidization velocity, U_{mf}, and the minimum velocity at which all the nozzles operate, U_m. Thus:

$$\begin{aligned} &\text{For} \quad \frac{U_m}{U_{mf}} \sim 1\text{–}2, \ \Delta p_D = 0.15\Delta p_B \\ &\text{and for} \quad \frac{U_m}{U_{mf}} \gg 1, \ \Delta p_D = 0.015\Delta p_B \end{aligned} \tag{1.100}$$

Zuiderweg's proposals were based on the measurements of Hiby[174] and of Whitehead and Dent[111], and he concluded that in most cases a conservative value for Δp_D of $0.10\Delta p_B$ should be adequate. The question of the optimum grid pressure drop is still open to dispute but it would seem prudent in practice to employ wherever possible a high rather than a low value. A high value entails greater expense in terms of gas-pumping costs but the risk of damage to the system caused by defluidized zones in the bed arising from poor gas distribution can be very great.

In addition to providing for uniform fluidization the distributor should be designed so as to minimize both attrition of the bed particles and erosion of its own structure and to prevent solids leaking into the windbox during operation and periods when the bed is deliberately defluidized for maintainance, repair, etc. This leaking of particles, also called 'dumping' or 'weeping', can be a particularly serious problem with simple perforated plate designs; any solid particles that enter the windbox can be entrained by the fluidizing gas and cause the grid to be eroded from below. Bergougnou and his co-workers have carried out a number of studies into the problem of grid leakage[175,176]. Based on the principle that the gas velocity through holes in a perforated plate distributor, U_g, can be expressed in terms of the flow of fluid through an orifice,

$$U_g = C_D \left[\frac{2(p_{WB} - p_G)}{\rho_g} \right]^{1/2} \tag{1.101}$$

where p_G is the gas pressure at the top of the hole, p_{WB} the gas pressure in the windbox, ρ_g the density and C_D an empirical discharge coefficient, they developed a simple model and tested it using grids of simulated varying thicknesses. They carried out experiments by placing a pressure transducer above a grid hole and measuring the pressure fluctuations. If fluctuations occur during which p_G exceeds p_{WB} then leakage may result depending on the frequency of the fluctuations and the thickness of the grid. It was found that such fluctuations in pressure did occur with a frequency corresponding to the frequency of bubble formation but that by selecting the proper thickness of grid the rate of leakage could be reduced by up to four orders of magnitude.

The work of Bergougnou *et al.* on grid leakage would seem to indicate that when gas enters a fluidized bed from the distributor it immediately forms into bubbles and that the frequency of bubble formation is more-or-less constant over a range of operating conditions. This is at variance with a good deal of thinking reported in the literature which assumes that gas enters as a jet that penetrates for some distance into the bed before breaking up into bubbles[177,178]. Belief in the existence of grid jets has been strengthened by visual observation of two-dimensional and semi-cylindrical beds where a stable flame-like structure can be seen issuing from an orifice into the bed[172,179]. However, using X-rays to examine gas flow from an orifice into a cylindrical bed Rowe *et al.*[180] found that jets are seldom if ever produced and that bubble formation occurs immediately on entry. They concluded that the only conditions under which a permanent jet is established over an orifice are if either the particles are not fluidized locally or there are surfaces near to the orifice on which a jet can stabilize. Rowe also found, in agreement with Bergougnou *et al.*, that the frequency of bubble formation remained roughly constant at about 8 Hz, an increase in gas flowrate causing the size of a bubble at detachment to increase.

Rowe further observed that on entry to the bed the bubble volume was usually only half that of the gas injected in the time; after detachment the bubble volume increased rapidly until, at a distance approximately two bubble diameters above the orifice, it had attained the expected value[181]. A qualitative explanation for this behaviour is that at the moment of its emergence from the orifice an element of gas flows at high velocity through the assembly of particles covering the hole and in so doing exerts a drag force on them. This causes the particles to move aside leaving a void space for the rest of the incoming gas to fill. Meanwhile the first gas to enter has penetrated some distance into the bed causing the local voidage to be increased above the stable value for interstitial flow; it is nevertheless in intimate contact with particles and can, in the case of a chemical reactor, undergo efficient conversion. The void at the orifice continues to grow

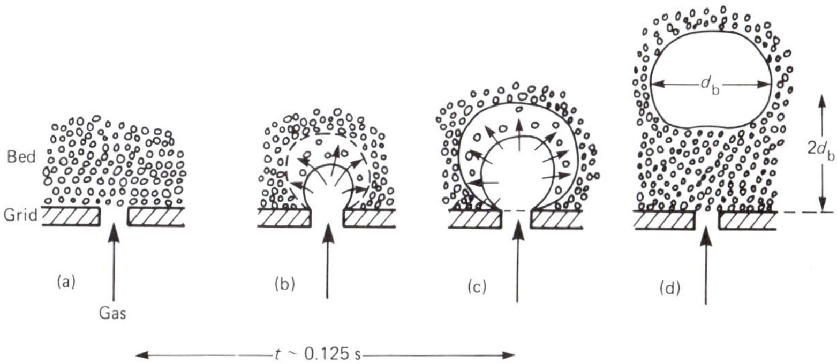

Figure 1.31 Bubble formation at an orifice (After Killick[182])
(a) gas enters bed of voidage ε_{mf}
(b) gas flows past particles, increasing their voidage to ε'_{mf} and forming incipient bubble
(c) incipient bubble of volume 0.5 V_b detaches from orifice surrounded by region of unstable voidage ε'_{mf}
(d) bubble of volume V_b fully formed at height $2d_b$ above grid and surrounded by dense phase at ε_{mf}

until the point of detachment is reached when it rises, comprising only half the gas that entered initially, into the bed. As this incipient bubble rises the 'surplus' gas in the dense phase flows back into it, probably through the base, until at a distance of two bubble diameters above the orifice the gas transfer is complete, the bubble is fully formed and the dense phase voidage has returned to its stable value. The sequence of events is shown diagrammatically in *Figure 1.31* and is referred to again in the context of a chemical reactor in Section 2.5.1.

Zenz used a concept very similar in practice to that suggested by the experimental evidence just discussed, to obtain a relationship between the hole diameter, D_o, and the hole spacing, x, of a perforated plate distributor[172]. Zenz assumed that stable gas jets equal in length to two bubble diameters, $2d_b$, are formed at the orifice and that the bubbles

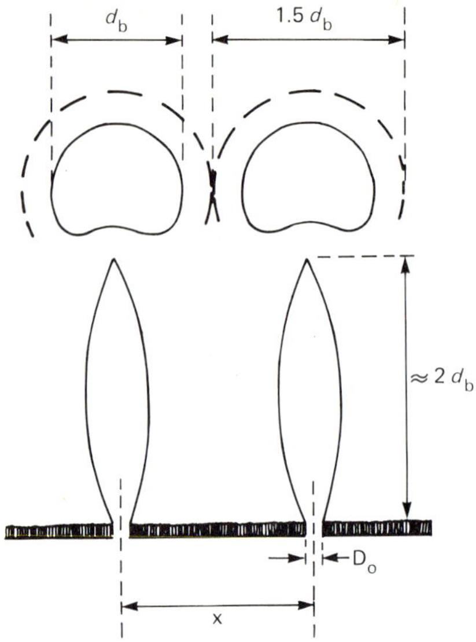

Figure 1.32 Idealized jet-to-bubble emergence pattern (From Zenz[172])

formed subsequently are surrounded by a shell of downward flowing particles of diameter $1.5d_b$ (*Figure 1.32*). He then argued that to prevent the premature growth of bubbles by coalescence, the separation, x should be a minimum of $1.5d_b$ and on the basis of experimental evidence from a number of studies, proposed the following correlation for a distributor pressure drop of $0.30\Delta p_B$ in a bed containing no restricting internal structures:

$$\log[3.52(H_f\rho_B)] = 1.3 + 0.0192\left(\frac{x}{D_o}\right) \quad (1.102)$$

where H_f is the height of the fluidized bed and ρ_B is the bed density. The correlation leads to the values given in *Table 1.3*.

Table 1.3 Minimum flat-plate grid hole spacings from equation 1.102 as a function of bed resistance (from Zenz[172])

Bed pressure drop $H_f\rho_B$	*Hole spacing/hole diameter* x/D_o
100	13
250	22.9
500	31.2

Although it would appear to be based on an erroneous physical model of gas entry, the Zenz correlation has been found to agree well with experimental data obtained using a 1 m diameter bed[169].

References

1 KOZENY, J., *Ber. Wien Akad.*, **136a**, 271 (1927)
2 CARMAN, P.C., *Trans. I. Chem. Engrs*, **15**, 150 (1937)
3 ERGUN, S., *Chem. Engng Prog.*, **48**(2), 89 (1952
4 WEN, C.Y. and YU, Y.H., *A.I.Ch.E. J*, **12**, 610 (1966)
5 GELDART, D., *Powder Tech.*, **7**, 285 (1973)
6 MASSIMILLA, L. and DONSI, G., *Powder Tech.*, **15**, 253 (1976)
7 McFARLANE, J.S. and TABOR, D., *Proc. R. Soc.*, **A202**, 224 (1950)
8 MUTSERS, S.M.P., PhD Thesis, Eindhoven University (1977)
9 RIETEMA, K. and MUTSERS, S.M.P., in *Fluidization*, Ed. Davidson, J.F. and Keairns, D.L., 81, Cambridge University Press (1978)
10 GUEDES DE CARVALHO, J.F.R. and HARRISON, D., *Inst. of Fuel Symp.* Ser. No.1, paper B1 (1975)
11 STEWART, P.B.S. and DAVIDSON, J.F., *Powder Tech.*, **1**, 61 (1967)
12 LEUNG, L.S., PhD Thesis, Cambridge University (1961)
13 HANCOCK, R.T., *Trans. Inst. Mining Engrs*, **94**, 114 (1937–38)
14 STEINOUR, H., *Ind. Engng Chem.*, **36**, 618 (1944)
15 LEWIS, W.K., GILLILAND, E.R. and BAUER, W.C., *Ind. Engng Chem.*, **41**, 1104 (1949)
16 LEWIS, E.W. and BOWERMAN, E.W., *Chem. Engng Prog.*, **48**, 603 (1952)
17 RICHARDSON, J.F. and ZAKI, W.N., *Trans. I. Chem. Engrs*, **32**, 35 (1954)
18 RICHARDSON, J.F., in *Fluidization*, Ed. Davidson, J.F. and Harrison, D., chapter 2, Academic Press, London, (1971)
19 ROWE, P.N., in *Fluidization*, Ed. Davidson, J.F. and Harrison, D., Chapter 4, Academic Press, London, (1971)
20 TWOOMY, R.D. and JOHNSTONE, H.F., *Chem. Engng Prog.*, **48**, 220 (1952)
21 GRACE, J.R. and CLIFT, R., *Chem. Engng Sci.*, **29**, 327 (1974)
22 LOCKETT, M.J., DAVIDSON, J.F. and HARRISON, D., *Chem. Engng Sci.*, **22**, 1059 (1967)
23 GRACE, J.R. AND HARRISON, D., *Chem. Engng Sci.*, **24**, 497 (1969)
24 PYLE, D.L. and HARRISON, D., *Chem. Engng Sci.*, **22**, 1199 (1967)
25 GODARD, K.E. and RICHARDSON, J.F., *Chem. Engng Sci.*, **23**, 660 (1968)
26 ROWE, P.N. and EVERETT, D.J., *Trans. I. Chem. Engrs*, **50**, 55 (1972)
27 ROWE, P.N. and YACONO, C.X.R., *Chem. Engng Sci.*, **31**, 1179 (1976)
28 ROWE, P.N., SANTORO, L. and YATES, J.G., *Chem. Engng Sci.*, **33**, 133 (1978)
29 DAVIDSON, J.F., *Trans. I. Chem. Engrs*, **39**, 230 (1961)
30 DAVIDSON, J.F. and HARRISON, D., *Fluidised Particles*, Cambridge University Press (1963)
31 KUNII, D. and LEVENSPIEL, O., *Fluidization Engineering*, Wiley, New York (1969)
32 REUTER, H., *Chem. Engng Prog. Symp. Ser.*, **62**, (62), 94 (1966)
33 ROWE, P.N., *Chem. Engng Prog.*, **60**(3), 75 (1964)
34 ROWE, P.N. and PARTRIDGE, B.A., *Trans. I. Chem. Engrs*, **43**, 157 (1965)
35 JACKSON, R., *Trans. I. Chem. Engrs*, **41**, 13,22 (1963)
36 JACKSON, R., in *Fluidization*, Ed. Davidson, J.F. and Harrison, D., Chapter 3, Academic Press, London, (1971)
37 MURRAY, J.D., *Chem. Engng Prog. Symp. Ser.*, **62**, (62), 57 (1966)
38 MURRAY, J.D., *J. Fluid Mech.*, **22**, 57 (1965)
39 COLLINS, R., *Chem. Engng Sci.*, **20**, 788 (1965)
40 COLLINS, R., in *Fluidization*, Ed. Grace, J.R. and Matsen, J.M., 117, Plenum Press, New York (1980)
41 DAVIES, R.M. and TAYLOR, G.I., *Proc. R. Soc.*, **A200**, 375 (1950)
42 DAVIDSON, J.F., HARRISON, D., DARTON, R.C. and LA NAUZE, R.D., in *Chemical Reactor Theory*, Ed. Lapidus, I. and Amundson, N.R., Chapter 10, Prentice-Hall, New York, (1977)
43 PYLE, D.L. and HARRISON, D., *Chem. Engng Sci.*, **22**, 531 (1967)
44 CHIBA, T., TERASHIMA, K. and KOBAYASHI, H., *Chem. Engng Sci.*, **27**, 965 (1972)
45 KOBAYASHI, H., ARAI, F. and CHIBA, T., *Kagaku Kogaku*, **4**(1), 147 (1966)
46 MOROOKA, S., TAJIMA, K. and MIYAUCHI, T., *Int. Chem. Engng*, **12**(1), 168 (1972)

47 WHITEHEAD, A.B., DENT, D.C. and BHAT, G.N., *Powder Tech.*, **1**, 143 (1967)
48 GRACE, J.R. and HARRISON, D., *Chem. Engng Sci.*, **22**, 1337 (1967)
49 GRACE, J.R. and VENTA, J., *Can. J. Chem. Engng*, **51**, 110 (1973)
50 SIT, S.P. and GRACE, J.R., *Chem. Engng Sci.*, **36**, 327 (1981)
51 WHITEHEAD, A.B. and YOUNG, A.D., *Proc. Int. Symp. Fluidization*, Eindhoven, 284, Netherlands University Press, Amsterdam, (1967)
52 GELDART, D., *Powder Tech.*, **6**, 201 (1972)
53 GELDART, D., *Powder Tech.*, **4**, 41 (1970/71)
54 WERTHER, J., *Trans. I. Chem. Engrs*, **52**, 149,160 (1974)
55 MASSON, H. and JOTTRAND, R., in *Fluidization*, Ed. Davidson, J.F. and Keairns, D.L., 1, Cambridge University Press (1978)
56 YASUI, G. and JOHANSON, L.N., A.I.Ch.E.Jl, **4**, 445 (1958)
57 ROWE, P.N., *Chem. Engng Sci.*, **31**, 285 (1976)
58 ROWE, P.N. and EVERETT, D.J., *Trans. I. Chem. Engrs*, **50**, 42 (1972)
59 YACONO, C.X.R., PhD Thesis, London University (1975)
60 DARTON, R.C., LA NAUZE, R.D., DAVIDSON, J.F. and HARRISON, D., *Trans. I. Chem. Engrs*, **55**, 274 (1977)
61 HARRISON, D. and LEUNG, L.S., in *Symposium on the interaction between fluids and solids*, Ed. Rottenburg, P.A., (127), Institution of Chemical Engineers (1962)
62 MATSEN, J.M., *A.I.Ch.E. Symp. Ser.*, **69**(128), 30 (1973)
63 HARRISON, D., DAVIDSON, J.F. and DE KOCK, J.W., *Trans. I. Chem. Engrs*, **39**, 202 (1961)
64 VAN SWAAIJ, W.P.M. and ZUIDERWEG, F.J., *Proc. 2nd Int. Symp. on Reaction Enging*, Amsterdam, B9–25 (1972)
65 UPSON, P.C. and PYLE, D.L., *Proc. Int. Conf. Fluidization and its Applications*, Toulouse, 207 (1973)
66 CLIFT, R., GRACE, J.R. and WEBER, M.E., *Ind. Engng Chem. Funds.*, **13**, 45 (1974)
67 RICE, W.J. and WILHELM, R.H., *A.I.Ch.E. Jl*, **4**, 423 (1958)
68 BELLMAN, R. and PENNINGTON, R.H., *Quart.Appl.Maths.*, **12**, 151 (1954)
69 ROWE, P.N., in *Fluidization*, Eds. Davidson, J.F. and Harrison, D., 151, Academic Press, London, (1971)
70 TAYLOR, G.I., Proc. R. Soc., **A201**, 192 (1950)
71 SCHUGERL, K., in *Fluidization*, Eds. Davidson, J.F. and Harrison, D., Chapter 6, Academic Press, London, (1971)
72 GELDART, D. and CRANFIELD, R.R., *Chem. Engng J.*, **3**, 211 (1972)
73 CRANFIELD, R.R. and GELDART, D., *Chem.Engng. Sci.*, **29**, 935 (1974)
74 PARK, D., LEVENSPIEL, O. and FITZGERALD, T.J., *Chem. Engng Sci.*, **35**, 295 (1980)
75 RIOS, G.M., BAXERRES, J.L. and GIBERT, H., in *Fluidization*, Eds. Grace, J.R. and Matsen, J.M., 529, Plenum Press, New York (1980)
76 MATHUR, K.B. and EPSTEIN, N., *Spouted Beds*, Academic Press, New York (1974)
77 BAKER, C.G.J. and GELDART, D., *Powder Tech.*, **19**, 177 (1978)
78 STEWART, P.B.S. and DAVIDSON, J.F., *Powder Tech.*, **1**, 61 (1967)
79 BAEYENS, J. and GELDART, D., *Chem. Engng Sci.*, **29**, 255 (1974)
80 MATSEN, J.M. and TARMY, B.L., *Chem. Engng Prog. Symp. Ser.*, **66**(101), 1 (1970)
81 GELDART, D., *Chem. and Ind.*, 1474 (1967)
82 YATES, J.G. and GREGOIRE, J–Y., *Chem. Engng Sci.*, **35**, 380 (1980)
83 YERUSHALMI, J., CANKURT, N.T., GELDART, D. and LISS, B., *A.I.Ch.E. Symp. Ser.*, **74**(176), 1 (1978)
84 YERUSHALMI, J., TURNER, D.H. and SQUIRES, A.M., *Ind. Engng Chem. Proc. Des. Dev.*, **15**, 47 (1976)
85 BAKKER, P.J. and HEERTJES, P.M., *Chem.Engng Sci.*, **12**, 260 (1960)
86 ZENZ, F.A. and WEIL, N.A., *A.I.Ch.E. Jl*, **4**, 472 (1958)
87 LEWIS, W.K., GILLILAND, E.R. and LANG, P.M., *Chem. Engng Prog. Symp. Ser.*, **58**(38), 65 (1962)
88 HEYWOOD, H., in *Symposium on the interaction between fluids and solids*, Ed. Rottenburg, P.A., 1, Institution of Chemical Engineers (1962)
89 PINCHBECK, P.H. and POPPER, F., *Chem. Engng Sci.*, **6**, 57 (1956)
90 LEVA, M. and WEN, C.Y., in *Fluidization*, Eds. Davidson, J.F. and Harrison, D., Chapter 14, Academic Press, London
91 YAGI, S. and AOCHI, T., Paper presented at Soc. Chem. Engng Japan, Spring Meeting (1955)

92 WEN, C.Y. and HASHINGER, R.F., *A.I.Ch.E. Jl*, **6**, 220 (1960)
93 MERRICK, D. and HIGHLEY, J., *A.I.Ch.E. Symp. Ser.*, **70**(137), 366 (1974)
94 MATSEN, J.M., *NSF Workshop on Research Needs and Priorities*, Rensselaer Poly. Inst., New York (October 1979)
95 GUGNONI, R.J. and ZENZ, F.A., in *Fluidization*, Eds. Grace, J.R. and Matsen, J.M., 501, Plenum Press, New York (1980)
96 GEORGE, S.E. and GRACE, J.R., *A.I.Ch.E. Symp. Ser.*, **74**(176), 67 (1978)
97 GELDART, D., CULLINAN, J., GEORGHIADES, S., GILVRAY, D. and POPE, D.J., *Trans. I. Chem. Engrs*, **57**, 269 (1979)
98 DO, H., GRACE, J.R. and CLIFT, R., *Powder Tech.*, **6**, 195 (1972)
99 KEHOE, P.W.K., PhD Thesis, Cambridge University (1969)
100 MORI, S. and WEN, C.Y., *A.I.Ch.E. Jl*, **11**, 109 (1975)
101 GELDART, D., *Powder Tech.*, **6**, 163 (1972)
102 WERTHER, J., in *Fluidization Technology*, Ed. Keairns, D.L., **1**, 16, Hemisphere Publishing Co., Washington, DC (1976)
103 WEN, C.Y. and CHEN, L.H., *A.I.Ch.E. Jl*, **28**, 117 (1982)
104 LARGE, J.F., MARTINIE, Y. and BERGOUGNOU, M.A., *Int. Conf. Powder and Bulk Solids Handling* (May 1976)
105 ROWE, P.N., PARTRIDGE, B.A., CHEYNEY, A.G., HENWOOD, G.A. and LYALL, E., *Trans. I. Chem. Engrs*, **43**, 271 (1965)
106 LIGHTHILL, M.J., *J. Fluid Mech.*, **1**, 31 (1956)
107 ROWE, P.N., *Chem. Engng Sci.*, **28**, 979 (1973)
108 BAEYENS, J. and GELDART, D., *Proc. Int. Conf. Fluidization and its Applications*, Toulouse, France, 182 (1973)
109 MARSHECK, R.M. and GOMEZPLATA, A., *A.I.Ch.E. Jl*, **11**, 167 (1965)
110 GABOR, J., *Chem. Engng Sci.*, **26**, 1247 (1971)
111 WHITEHEAD, A.B. and DENT, D.C., *Proc. Int. Symp. Fluidization*, Eindhoven, 802, Netherlands University Press, Amsterdam (1967)
112 WHITEHEAD, A.B. and YOUNG, A.D., *Powder Tech.*, **1**, 149 (1967)
113 WHITEHEAD, A.B., GARTSIDE, G. and DENT, D.C., *Powder Tech.*, **14**, 61 (1976)
114 NGUYEN, H.V., WHITEHEAD, A.B. and POTTER, O.E., *A.I.Ch.E. Jl*, **23**, 913 (1977)
115 SINGH, B., RIGBY, G.R. and CALLCOTT, T.G., *Inst. of Fuel Symp. Ser. No. 1*, paper C5 (1975)
116 NGUYEN, H.V., WHITEHEAD, A.B. and POTTER, O.E., in *Fluidization*, Eds. Davidson, J.F. and Keairns, D.L. 140, Cambridge University Press (1978)
117 WHITEHEAD, A.B., POTTER, O.E. and NGUYEN, H.V., in *Fluidization*, Eds. Grace, J.R. and Matsen, J.M., 333, Plenum Press, New York (1980)
118 WERTHER, J., *Powder Tech*, **15**, 155 (1976)
119 SCHMALFELD, J., *Dissertation*, Karlsruhe University (1973)
120 MERRY, J.M.D. and DAVIDSON, J.F., *Trans. I. Chem. Engrs*, **51**, 361 (1973)
121 MERRY, J.M.D. and DAVIDSON, J.F., *Proc. Int. Conf. Fluidization and its Applications*, 196, Toulouse, France, (1973)
122 ROWE, P.N., NIENOW, A.W. and AGBIM, A.J., *Trans. I. Chem. Engrs*, **50**, 310 (1972)
123 NIENOW, A.W., ROWE, P.N. and CHEUNG, L.Y.L., in *Fluidization*, Eds. Davidson, J.F. and Keairns, D.L., 146, Cambridge University Press (1978)
124 ROWE, P.N. and NIENOW, A.W., *Chem. Engng Sci.*, **30**, 1365 (1975)
125 GIBILARO, L.G. and ROWE, P.N., *Chem. Engng Sci.*, **29**, 1403 (1974)
126 NIENOW, A.W. and CHEESMAN, D.J., in *Fluidization*, Eds. Grace, J.R. and Matsen, J.M., 373, Plenum Press, New York (1980)
127 GRACE, J.R., *Can. J. Chem. Engng*, **48**, 30 (1970)
128 UNO, S. and KITNER, R.C., *A.I.Ch.E. Jl*, **2**, 420 (1956)
129 ANGELINO, H., *Chem. Engng Sci.*, **21**, 541 (1966)
130 MENDELSON, H.D., *A.I.Ch.E. Jl*, **13**, 250 (1967)
131 ANGELINO, H., CHARZAT, C. and WILLIAMS, J.R., *Chem. Engng Sci.*, **19**, 289 (1964)
132 DAVENPORT, W.G., BRADSHAW, A.V. and RICHARDSON, F.D., *J. Iron Steel Inst.*, **205**, 1034 (1967)
133 HABERMAN, W.L. and MORTON, R.K., *Trans. Am.Soc. Civil Engrs*, **121**, 227 (1956)

134 SLAUGHTER, I. and WRAITH, A.E., *Chem. Engng Sci.*, **23**, 932 (1968)
135 BATCHELOR, G.K., *An Introduction to Fluid Mechanics*, Cambridge University Press (1967)
136 TOEI, R, MATSUNO, R., KOJIMA, H., NAGAI, Y., NAKAGAWA, K. and YU, S., *Chem. Engng Japan*, **4**, 142 (1966)
137 FRANKEL, N.A. and ACRIVOS, A., *Chem. Engng Sci.*, **22**, 847 (1967)
138 HOFFERT, F.D., KELLY, E.A. and SQUIRES, A.M., *Mech. Engr*, **81**, 27 (1959)
139 LEE, B.S., PYRCIOCH, E.J. and SCHORA, F.C. *Chem. Engng Prog. Symp. Ser.*, **66**(105), 152 (1970)
140 BOTTERILL, J.S.M. and DESAI, M., *Powder Tech.*, **6**, 231 (1972)
141 GODARD, K. and RICHARDSON, J.F., *Proc. Tripartite Chem. Engng Conf.*, Montreal, 126 (1968)
142 ZENZ, F.A., *Ind. Engng Chem. Funds.*, **3**, 65 (1964)
143 GUEDES DE CARVALHO, J.R.F., KING, D.F. and HARRISON, D., in *Fluidization*, Eds. Davidson, J.F. and Keairns, D.L., 59, Cambridge University Press (1978)
144 KING, D.F. and HARRISON, D., in *Fluidization*, Eds. Grace, J.R. and Matsen, J.M., 101, Plenum Press (1980)
145 SUBZWARI, M.P., CLIFT, R. and PYLE, D.L., in *Fluidization*, Eds. Davidson, J.F. and Keairns, D.L., 50, Cambridge University Press, (1978)
146 ROWE, P.N. and MACGILLIVRAY, H.J., *Inst. of Energy Symp. Ser. No. 4*, Paper 4.1, London (1980)
147 KAWABATA, J., YUMIYAMA, M., TAZAKI, Y., HONMA, S., CHIBA, T., SUMIYA, T. and ENDO, K., *J. Chem. Engng Japan*, **14**, 85 (1981)
148 ROWE, P.N., YATES, J.G., FOSCOLO, P. and HOFFMANN, A.C., *Chem. Engng Sci.*, **37**, 1115 (1982)
149 BOTTERILL, J.S.M. *Fluid-bed Heat Transfer*, Academic Press, London (1975)
150 LEVENSPIEL, O. and WALTON, J.S., *Chem. Engng Prog. Symp. Ser.*, **50**(9), 1 (1954)
151 WEN, C.Y. and LEVA, M., *A.I.Ch.E. Jl*, **2**, 482 (1956)
152 WENDER, I. and COOPER, G.T., *A.I.Ch.E. Jl*, **4**, 15 (1958)
153 VREEDENBERG, H.A., *Chem. Engng Sci.*, **9**, 52 (1958); **11**, 274 (1960)
154 MICKLEY, H.S. and TRILLING, C.A., *Ind. Engng Chem.*, **41**, 1135 (1949)
155 MICKLEY, H.S. and FAIRBANKS, D.F., *A.I.Ch.E. Jl*, **1**, 374 (1955)
156 MICKLEY, H.S., FAIRBANKS, D.F. and HAWTHORN, R.D., *Chem. Engng, Prog. Symp. Ser.*, **57**(32), 51 (1961)
157 BASKAKOV, A.P., *Int. Chem. Engng*, **4**, 220 (1964)
158 GELPERIN, N.I. and EINSTEIN, V.G., in *Fluidization*, Eds. Davidson, J.F. and Harrison, D., Chapter 10, Academic Press, London (1971)
159 BOTTERILL, J.S.M. and WILLIAMS, J.R., *Trans. I. Chem. Engrs*, **41**, 217 (1963)
160 BOTTERILL, J.S.M., BUTT, M.H.D., CAIN, G.L. and REDDISH, K.A., *Proc. Int. Conf. Fluidization*, Eindhoven, 442, Netherlands University Press, Amsterdam (1967)
161 COUDERC, J.P., ANGELINO, H., ENJALBERT, M. and GUIGLION, C., *Chem. Engng Sci.*, **22**, 99 (1967)
162 GABOR, J., *Chem. Engng Prog. Symp. Ser.*, **66**(105), 76 (1970)
163 XAVIER, A.M. and DAVIDSON, J.F., in *Fluidization*, Eds. Davidson, J.F. and Keairns, D.L., 333, Cambridge University Press (1978)
164 YAGI, S. and KUNII, D., *A.I.Ch.E. Jl*, **3**, 373 (1967)
165 YAGI, S., KUNII, D. and WAKAO, N., *Int. Dev. Heat Trans.*, Part 4, 742 (1961)
166 KUNII, D. and SMITH, J.M., *A.I.Ch.E. Jl*, **6**, 71 (1960)
167 GABOR, J., *Chem. Engng Sci.*, **25**, 979 (1970)
168 XAVIER, A.M., KING, D.F., DAVIDSON, J.F. and HARRISON, D., in *Fluidization*, Eds. Grace, J.R. and Matsen, J.M., 209, Plenum Press, New York (1980)
169 WERTHER, J., in *Fluidization*, Eds. Davidson, J.F. and Keairns, D.L., 7, Cambridge University Press (1978)
170 PETRIE, J.C. and BLACK, D.E., *Chem. Engng Prog.*, **62**(3), 82 (1966)
171 AGARWAL, J.C., DAVIS, W.L. and KING, D.T., *Chem. Engng Prog.*, **58**(11), 85 (1962)
172 ZENZ, F.A., *Proc. Tripartite Conf. Fluidization*, 136, Montreal, (1968)
173 ZUIDERWEG, F.J., *Proc. Int. Conf. Fluidization*, 739, Eindhoven; Netherlands University Press, Amsterdam (1967)
174 HIBY, J.W., *Chem. Engng Techn.*, **36**, 228 (1964)
175 BEHIE, L.A., VOEGELIN, B.E. and BERGOUGNOU, M.A., *Can.J. Chem. Engng*, **56**, 404 (1978)

176 BRIENS, C., BERGOUGNOU, M.A. and BAKER, C.G.J., in *Fluidization*, Eds. Grace, J.R. and Matsen, J.M., 413, Plenum Press, New York (1980)
177 MERRY, J.M.D., *Trans. I. Chem. Engrs*, **49**, 189 (1971)
178 MARKHEVA, V.I., BASOV, V.A., MELIK-AKHNAZAROV, T.K. and OROCHKO, D.I., *Theor. Found. Chem. Engng*, **5**, 80 (1971)
179 YANG, W.C. and KEAIRNS, D.L., in *Fluidization*, Eds. Davidson, J.F. and Keairns, D.L., 208, Cambridge Univesity Press (1978)
180 ROWE, P.N., MACGILLIVRAY, H.J. and CHEESMAN, D.J., *Trans. I. Chem. Engrs*, **57**, 194 (1979)
181 ROWE, P.N., private communication
182 KILLICK, R.C., PhD Thesis, London University (1982)

CHAPTER 2

Fluidized-bed reactor models

2.1 Introduction

A simple example will illustrate some of the considerations involved in the design of a fluid-bed reactor. Consider a reactor in which benzene vapour at 350°C is to be oxidized catalytically with excess air to maleic anhydride:

$$C_6H_5 + 4\tfrac{1}{2}O_2 = C_4H_2O_3 + CO_2 + H_2O; \Delta H = -468\,\text{kJ/mol}$$

The catalyst is available as 200 μm diameter particles which fluidize at 0.04 m/s at 350°C and over which the oxidation reaction is first order in benzene with a rate constant of $k\,s^{-1}$. Catalyst temperature is to be maintained constant at 350°C and the bed wall through which the exothermic heat of reaction is to be withdrawn at some lower temperature T_w°C. Reactor feed is air containing 4 per cent by volume of benzene (i.e. 10 per cent by mass benzene, 21 per cent oxygen and 69 per cent nitrogen); the air inlet pressure is 2.5 bar and the bed pressure drop 1 bar. It is required to produce 20 000 tonnes per 340 day year of maleic anhydride with a benzene conversion of 95 per cent.

From the stoichiometry of the reaction it is readily seen that the production required is equivalent to the conversion of 0.541 kg/s of benzene, which, assuming a conversion of only 95 per cent, gives a benzene flowrate of about 0.15 m^3/s at 350°C. The required air flowrate is thus 3.75 m^3/s. The minimum bed diameter is then found from the air flowrate and the maximum gas velocity consistent with maintaining the catalyst particles within the bed. Thus a maximum gas velocity of say 0.5 m/s ($12.5U_{mf}$) results in a minimum bed diameter of about 3 m. This value would of course need to be revised if heat transfer were to be via internal cooling coils.

Having obtained a value for bed diameter it is then necessary to calculate the minimum height needed to achieve a 95 per cent benzene conversion given the reaction kinetics and with information on the extent of contact between the reactant gases (assumed to be premixed) and the catalyst particles. Finally the wall area equivalent to this bed height must be shown to be compatible with the heat transfer requirements of the system.

At this stage the most crucial question is: how much contact is there between reactant and catalyst as the gas flows through the bed? To answer this requires the application of theories of two-phase flow in fluidized beds and the development of a satisfactory model to describe the flow. It is generally accepted that at gas velocities in excess of the minimum fluidization value the gas entering the bed divides into two streams, one flowing through the emulsion phase and the other flowing as bubbles (Section 1.6.1). Gas flowing in the emulsion phase is clearly in intimate contact with bed particles and so reaction between them can proceed efficiently. Bubbles however are essentially empty of particles and the the gas within them can only react at the bubble 'wall' and in the limited cloud-wake region surrounding the bubble. (With beds of large particles the cloud-wake does not form and the bubble merely acts as a short circuit for the gas flow.)

The matter is made more complicated by the fact that gas exchange occurs between the emulsion and bubble phases, the bubbles thereby acting as secondary distributors of fresh reactant as they rise through the bed. The performance of a fluid-bed reactor is thus determined by a combination of chemical factors (stoichiometry, kinetics, etc.) and hydrodynamic factors (gas distribution, bubble size and residence time, interphase exchange rate, etc.) and the precise way in which this combination is described quantitatively constitutes the reactor model.

The relative importance of the hydrodynamic and chemical factors has been demonstrated in an elementary way by Grace[1]. Assuming a simple first-order reaction of the type considered above, with a bed of height H fluidized at a superficial gas velocity U, a fraction fb of the incoming gas flowing in the bubble phase and $(1 - \beta)$ in the dense phase, then the fraction of reactant unconverted, $C_{out}/C_{in} = C'$, may be expressed in terms of a dimensionless reaction rate group, $k' = kH/U$, for the following simple models:

(a) a pseudo-homogeneous reactor with plug flow of reactant gas:

$$C' = e^{-k'} \tag{2.1}$$

(b) a pseudo-homogeneous perfectly mixed reactor:

$$C' = \frac{1}{1 + k'} \tag{2.2}$$

(c) a two-phase reactor with a perfectly mixed dense phase, no reaction in the bubble phase and applying the model of Davidson and Harrison[2] (see Section 2.3.2):

$$C' = f_b e^{-X} + \frac{(1 - f_b e^{-X})^2}{1 - f_b e^{-X} + k'} \tag{2.3}$$

where f_b is the fraction of bed occupied by bubbles and X is an inter-phase mass transfer parameter.

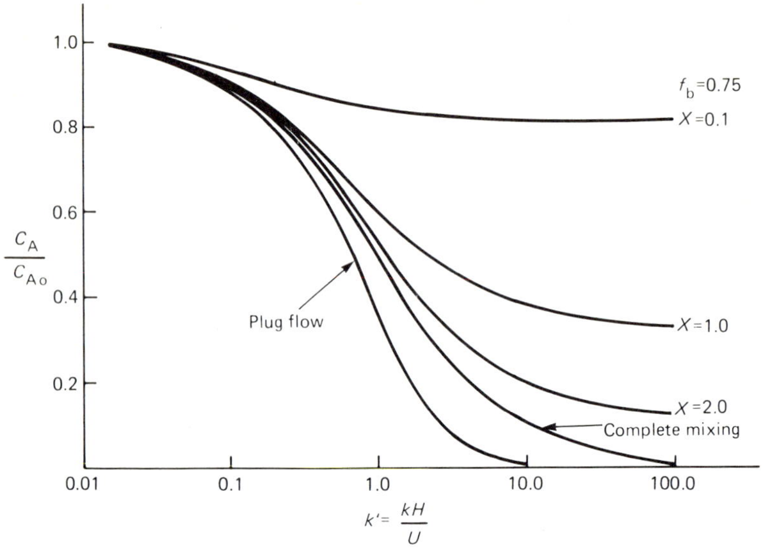

Figure 2.1 Fraction of reactant A, unconverted as a function of dimensionless rate coefficient for different models of contacting (After Grace[1])

Predicted conversions from the three models are plotted in *Figure 2.1* with, for model (c), an f_b value of 0.75 and a range of X values from 0.1 to 2.0. It is apparent that for low values of k' the hydrodynamics have only a minor influence on reactor performance and it makes little difference to the predictions which model is chosen. For higher k' values however, the hydrodynamics become of overriding importance and it is then vital to choose the correct model to obtain an accurate prediction of reactor performance.

2.2 General approach to reactor modelling

2.2.1 Introduction

The Davidson–Harrison model is one of a large number of fluidized-bed reactor models to have been developed in recent years. Most are based on the two-phase theory of fluidization but differ considerably in the assumptions they make regarding the exact nature of the phases, the mode of inter-phase gas exchange, the degree of gas mixing in the phases etc. A number of reviews of the models have been published[3,4,5,6] and in most cases it has been found convenient to divide them into two types: (i) simple or arbitrary models based on empirical correlations obtained with small-scale equipment, (ii) models based on bubble dynamics which describe reactor behaviour in terms of the known physics and hydrodynamics of fluid beds (*Figure 2.2*). A modified version of the comparisons by Grace[3] and Rowe[4] is given in *Table 2.1* and *2.2*.

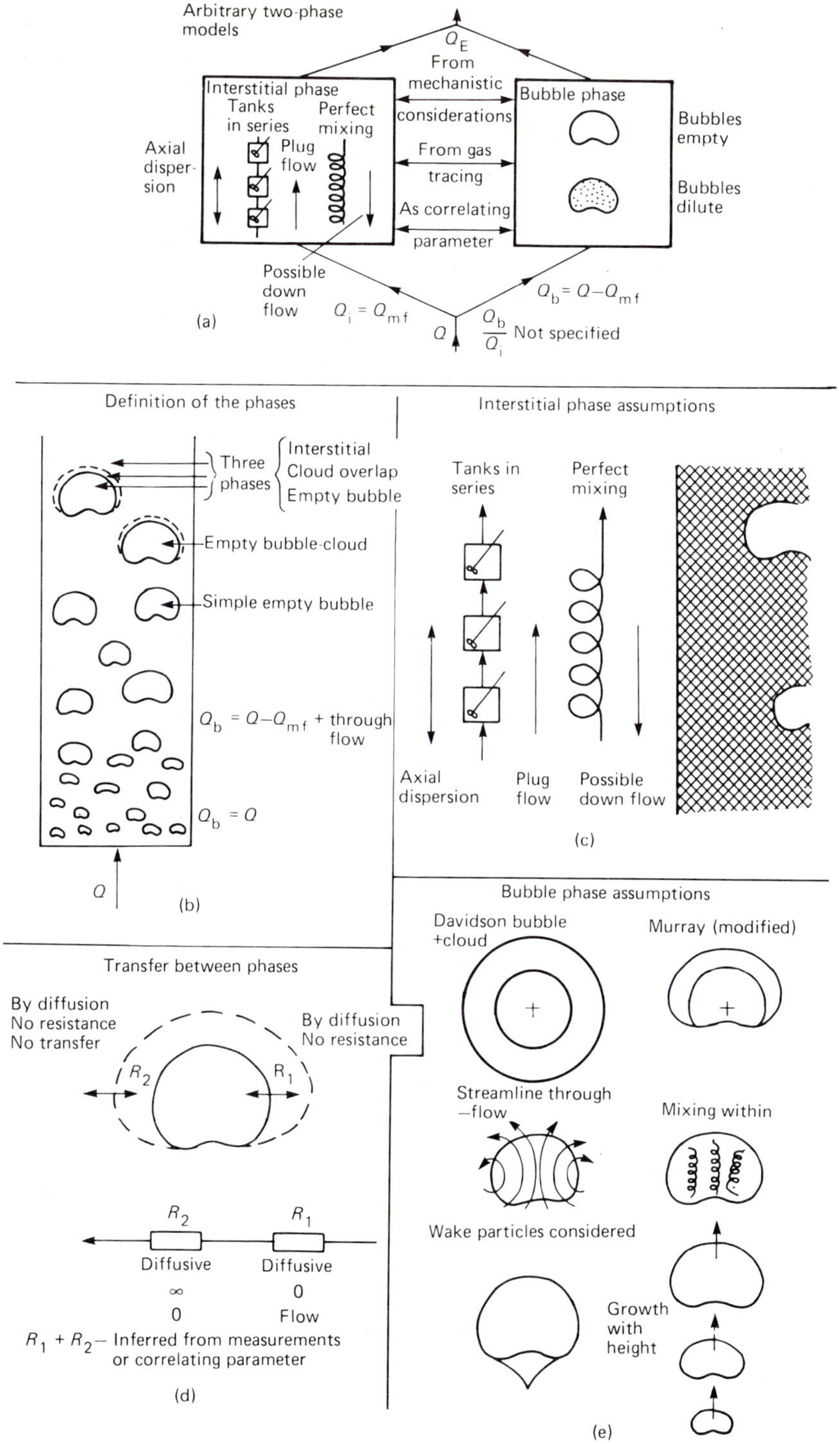

Figure 2.2 Fluid-bed reactor models (From Rowe[4])

Table 2.1

	Phase division		Bubble flow, Q_b			Emulsion flow					Interphase transfer		
	Two phases bubbles empty	Two phases, bubbles contain some solids	$Q_b = Q - Q_{mf}$	$Q_b = Q$	Not specified	Axial dispersion model	Tanks in series model	Plug flow	Completely mixed	Down flow allowed	Measured by gas mixing or kept as correlating parameter	Calculated from chemical rate data	Calculated mechanistically
Shen and Johnstone [7]	⊘		⊘					⊘	⊘			⊘	
Mathis and Watson [8]		⊘	⊘					⊘				⊘	
Lewis et al [9]		⊘		⊘				⊘	⊘			⊘	
May [10]	⊘		⊘			⊘					⊘		
Lanneau [11]		⊘			⊘					⊘			⊘
Gomezplata and Shuster [12]	⊘		⊘					⊘				⊘	
Van Deemter [13]	⊘				⊘	⊘					⊘		
Mamuro and Muchi [14]	⊘		⊘				⊘					⊘	
Kobayashi et al [15]		⊘	⊘	⊘				⊘	⊘		⊘		
Mireur and Bischoff [16]	⊘		⊘			⊘					⊘	⊘	
Van Deemter [17]	⊘				⊘			⊘		⊘	⊘		

Table 2.2

	Phase division			Bubble flow, Q_b		Emulsion flow						Interphase transfer							Cloud size		Flow in bubble		Flow in cloud		Wake	
	Two phases, bubbles empty	Two phases: cloud + bubble+ wake; emulsion	Three phases; bubble; cloud + wake; emulsion	$Q_b = Q - Q_{mf}$	$Q_b = Q$	Axial dispersion (1)	Axial dispersion (2)	Tanks in series	Plug flow	Completely mixed	Downflow allowed	(1)	(2)	(3)	(4)	(5)	(6)	(7)	Davidson	Murray	Streamline	Completely mixed	Streamline	Completely mixed	Included in cloud	Ignored or included in dense phase
Orcutt et al [18]	⊘			⊘					⊘	⊘					⊘	⊘						⊘				⊘
Davidson and Harrison [2]	⊘			⊘					⊘	⊘					⊘	⊘						⊘				⊘
Rowe [19]		⊘		⊘					⊘					⊘				⊘	⊘			⊘		⊘	⊘	
Rose [20]		⊘		⊘					⊘						⊘			⊘	⊘		⊘		⊘			
Partridge and Rowe [21]		⊘		⊘					⊘					⊘			⊘			⊘		⊘		⊘	⊘	
Toor and Calderbank [22]	⊘	⊘		⊘		⊘			⊘	⊘					⊘	⊘	⊘	⊘	⊘			⊘	⊘	⊘		⊘
Hovmand and Davidson [23]	⊘			⊘					⊘						⊘		⊘					⊘				⊘
Kunii and Levenspiel [24]			⊘	⊘			⊘				⊘				⊘		⊘		⊘			⊘		⊘	⊘	
Kato and Wen [25]		⊘			⊘			⊘				⊘		⊘					⊘			⊘		⊘		⊘
Chiba and Kobayashi [26]		⊘		⊘					⊘				⊘	⊘						⊘		⊘		⊘	⊘	
Fryer and Potter [27]			⊘	⊘			⊘			⊘	⊘				⊘		⊘		⊘			⊘		⊘	⊘	

Although we will not examine all the models in detail it is worthwhile considering some of the major points of difference among them. Most models accept the division of flow into two phases but the detailed description, particularly of the bubble phase, varies; this becomes important when the gas exchange mechanism is formulated (*see* below). Furthermore the majority of models accept the bubble flow as being all that in excess of the minimum fluidization value, but again, differences arise in their description of the flow, streamline or turbulent, inside the bubbles and clouds.

The two factors causing the greatest difference in the prediction of overall performance however are interphase transfer and gas mixing in the emulsion phase, and we will consider these a little more fully.

Notes on Table 2.2

(i) *Emulsion flow*
Axial dispersion (1): dispersion coefficient found from gas mixing study or correlated
Axial dispersion (2): found from chemical reaction study

(ii) *Interphase exchange*
(1) rate found from gas mixing study or correlated
(2) rate found from chemical reaction
(3) no resistance to bubble–cloud transfer
(4) bubble–cloud transfer by diffusion and/or throughflow
(5) no resistance to cloud–emulsion transfer
(6) cloud–emulsion transfer by diffusion
(7) no cloud–emulsion transfer

2.2.2 Interphase gas exchange

2.2.2.1 Single bubbles

Although there are some superficial similarities between gas bubbles in liquids and those in fluidized beds, a major difference lies in the presence in the former case of a relatively impervious surface layer of liquid around the bubble which is maintained in existence by surface tension forces. There is no coherent surface and no surface tension force surrounding a gas bubble in a fluidized bed and as a result it is possible for gas within a bubble to flow through the interstices between the particles and to exchange with gas in the emulsion phase of the bed. This exchange will have an important influence on the extent of gas–solid contact and hence on the level of chemical conversion that it may be possible to achieve under given conditions, and it represents a major factor in any model of a reacting fluidized bed system.

It has been pointed out[28] that there are three 'classical' approaches to

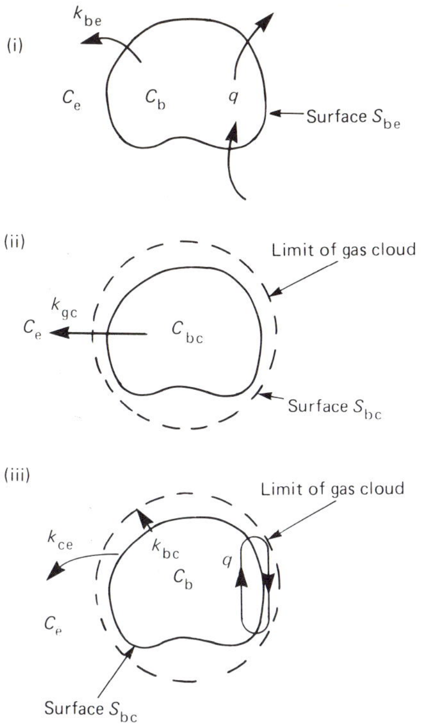

Figure 2.3 Models of interphase gas transfer; (i) direct bubble to emulsion, (ii) bubble-cloud to emulsion, (iii) bubble to cloud to emulsion

the problem of predicting mass transfer from single bubbles and all have been explored in reactor models. The three approaches will now be discussed.

(i) Direct bubble-to-emulsion transfer (*Figure 2.3(i)*). Historically this was the first to be tried and was developed in the studies of Zenz and Othmer[29], Orcutt *et al.*[18] and Davidson and Harrison[2]. Essentially the theory assumes the gas transfer from bubble to emulsion to be caused by a combination of convective throughflow, given by the term q, and molecular diffusion. The two effects are assumed to be additive and so the molar rate of gas transfer is given by:

$$\frac{-dV_bC_b}{dt} = (q + k_{be}S_{be})(C_b - C_e) \tag{2.4}$$

where k_{be} is a mass transfer coefficient, S_{be} is the interphase surface area and C_b and C_e are the concentrations of gas in the bubble and emulsion phases respectively. Now, from Davidson's theory[2] the throughflow term for a spherical bubble of diameter d_b is:

$$q = \frac{3\pi}{4} U_{mf} d_b^2 \tag{2.5}$$

while k_{be}, evaluated by analogy with film-controlled diffusion from a spherical-cap shaped gas bubble rising through a liquid, is:

$$k_{be} = 0.975 D_G^{1/2} d_b^{-1/4} g^{1/4} \qquad (2.6)$$

D_G being the molecular diffusivity of the gas. The overall exchange rate (volume of gas per unit volume of bubble per unit time) is then:

$$K_{be} = \frac{(q + k_{be}S_{be})}{V_b} \qquad (2.7)$$

$$= 4.5\left(\frac{U_{mf}}{d_b}\right) + 5.85\left(\frac{D_G^{1/2}g^{1/4}}{d_b^{5/4}}\right) \qquad (2.8)$$

Using the transfer equation 2.4, assuming the dense phase to be completely mixed and with no reaction in the bubble phase, Davidson and Harrison[2] obtained equation 2.3 for the fraction of reactant leaving unchanged from the bed surface. An essential feature of the model is the assumption of a uniform bubble size throughout the bed, although Calderbank and Toor[30] have extended the model to take account of the variation in bubble size observed in practice.

(ii) Cloud-to-emulsion transfer (*Figure 2.3(ii)*). Here the cloud surrounding the bubble is taken into account and the bubble–cloud phase is assumed to be perfectly mixed. Transfer then occurs primarily by diffusion across the cloud-emulsion boundary giving:

$$\frac{-d(V_{bc}C_{bc})}{dt} = k_{gc}S_{bc}(C_{bc} - C_e) \qquad (2.9)$$

V_{bc} being the volume of gas in the bubble and cloud, S_{bc} the cloud surface area and k_{gc} the cloud–emulsion mass transfer coefficient. Partridge and Rowe[21] by analogy with transfer from a drop of one immiscible liquid rising through another assumed the mass transfer coefficient to be given by the Sherwood number:

$$Sh_c = \frac{k_{gc}d_c}{D_G} = 2 + 0.69 Sc^{1/3} Re_c^{1/2} \qquad (2.10)$$

where Sc is the Schmidt number:

$$Sc = \frac{\mu}{\rho D_G} \qquad (2.11)$$

and Re_c is a Reynolds number defined in terms of the relative velocity, U_R, between rising cloud and emulsion ($U_R = U_c - U_e$)

$$Re_c = \frac{U_R d_c \rho}{\mu} \tag{2.12}$$

Here the symbols have their usual meanings, i.e. d_c is the diameter of the sphere with the same volume as the cloud, while D_G, μ and ρ represent diffusivity, viscosity and density respectively, of the fluidizing gas.

The cloud-to-emulsion exchange rate per unit volume of cloud space is then:

$$K_{bce} = \frac{k_{gc}\,\pi\, d_c^2\, \varepsilon}{V_{bc}} \tag{2.13}$$

$$= \frac{3.9\varepsilon D_G Sh_c}{V_{bc}^{2/3}} \tag{2.14}$$

ε being the emulsion-phase voidage. Equation 2.14 has been used in the models of Rowe[19] and Partridge and Rowe[21].

(iii) Bubble-to-cloud-to-emulsion transfer (*Figure 2.3(iii)*). This approach, developed by Kunii and Levenspiel[24], considers transfer to occur in two stages, (a) bubble-to-cloud-wake region expressed by the Davidson–Harrison equation 2.8, and (b) cloud-wake-to-emulsion given by an expression obtained from the Higbie penetration theory[31]:

$$K_{ce} = 6.78 \left(\frac{\varepsilon_{mf} D_G U_b}{d_b^3} \right)^{1/2} \tag{2.15}$$

The overall transfer rate, K_{be}, is then found from:

$$\frac{1}{K_{be}} = \frac{1}{K_{bc}} + \frac{1}{K_{ce}} \tag{2.16}$$

In most practical cases however, the transfer from cloud to emulsion is rate controlling:

$$K_{be} \simeq K_{ce}$$

and the model becomes very similar to the purely diffusive type described under approach (ii).

As part of a comparative study of reactor models (to be discussed later) Chavarie and Grace[28] measured mass transfer rates for bubbles containing ozone injected into an air-fluidized two-dimensional bed of particles. They were able both to eliminate entrance effects, which can cause enhanced mass transfer during bubble formation, and to allow for bubble growth in the bed. They compared their empirical mass transfer coefficient, k_{GT}, defined by:

$$\frac{-\mathrm{d}(V_b C_b)}{\mathrm{d}t} = k_{GT} S_b (C_b - C_e) \tag{2.17}$$

with the predictions of the models mentioned and concluded that those which assume a combination of throughflow and diffusion overestimate the rate of transfer, while those which assume a purely diffusive mechanism underestimate it (*Table 2.3*). The Kunii–Levenspiel model, being controlled by the diffusive term also leads to an underestimate of the rate. The authors concluded that an alternative approach based on the throughflow relationship of Murray[32,33] and considering the bubbles to be circular in shape, leads to expressions:

Table 2.3 Interphase mass transfer models and coefficients for ozone exchange in an air-fluidized two-dimensional bed (from Chavarie and Grace[28])

Model	*Reference*	*Coefficient*	k_{GT} *equivalence*	*Calculated* k_{GT} (m/s)
1 Partridge and Rowe	21	$k_{gc} = \dfrac{0.26 D_G Re_c^{0.6} Sc^{0.33}}{d_c}$	$\varepsilon_{mf} k_{GC}$	0.0020
2 Chiba and Kobayashi	26	$k_{gc} = \sqrt{\left[\dfrac{4 D_G \varepsilon_{mf}^2 U_b(\alpha - 1)}{\pi d_c \alpha}\right]}$	k_{GC}	0.0030
3 Davidson and Harrison	2	$k_{be} = 0.6 D_G^{1/2}\left(\dfrac{g}{d_b}\right)^{1/4}$ $q = 2U_{mf} d_b w$ $Q = q + k_{be} S_{be}$	Q/S_b	0.0410
4 Kunii and Levenspiel	24	$k_{bc} = 0.6 D_G^{1/2}\left(\dfrac{g}{d_b}\right)^{1/4}$ $q = 2U_{mf} d_b w$ $Q = q + k_{bc} S_{bc}$ $k_{ce} = \dfrac{4 D_G \varepsilon_{mf} U_b}{\pi d_c}$	$\dfrac{k_{ce} Q}{k_{ce} S_{bc} + Q}$	0.0051
5 Murray throughflow, *no diffusion*	32, 33	$k_{bc} = 0$ $q = U_{mf} d_b w$	q/S_{bc}	0.0160
Empirical value				0.016

$$Q = q = U_{mf}\, d_b w \tag{2.18}$$

(where w is the inside thickness of the two-dimensional bed), and

$$k_{GT} = \frac{U_{mf}}{\pi} \tag{2.19}$$

that are in good agreement with the measured value.

In deriving equation 2.19 the diffusive transfer was assumed to be swamped by the convective flow, although the case where diffusion continues to play a part was explored by Sit and Grace[34] who obtained an overall mass-transfer coefficient for a circular two-dimensional bubble:

$$k_{GT} = \frac{U_{mf}}{\pi} + \left(\frac{4D_G\,\varepsilon_{mf}\,U_b}{\pi d_b}\right)^{1/2} \tag{2.20a}$$

and for a spherical three-dimensional bubble:

$$k_{GT} = \frac{U_{mf}}{4} + \left(\frac{4D_G\,\varepsilon_{mf}\,U_b}{\pi d_b}\right)^{1/2} \tag{2.20b}$$

The first term on the right-hand side of these expressions represents the convective throughflow contribution while the second term is the diffusive component. The overall transfer rate per unit volume of bubble may be found by multiplying the mass transfer coefficients by the appropriate surface-to-volume ratio. Thus for a spherical bubble in a three-dimensional bed:

$$K_{GT} = \frac{6}{d_b}\,k_{GT} \tag{2.21a}$$

$$= \frac{1.5U_{mf}}{d_b} + \frac{12}{d_b^{3/2}}\left(\frac{D_G\varepsilon_{mf}U_b}{\pi}\right)^{1/2} \tag{2.21b}$$

2.2.2.2 *Multiple bubbles*

Although work on single bubbles in two-dimensional beds can lead to useful comparisons its conclusions should be treated with caution when applied to freely bubbling, three-dimensional systems. Here bubble splitting, coalescence, wake shedding, etc. occur as a matter of course and the interphase mass transfer mechanism is almost certain to be more complicated than any considered so far. This is borne out by the study of Sit and Grace[35], already referred to in Chapter 1 (Section 1.6.4), who investigated the transfer of ozone between pairs of bubbles and the emulsion phase during coalescence. They estimated the mass-transfer coefficient experienced by the leading bubble of a coalescing pair to be two to three times greater during the 'approach period' than when the bubble was in isolation. They also found the enhancement of interphase mass transfer to increase with particle size, a result in accord with a number of other studies (*Table 2.4*).

Now the throughflow term in the expressions for transfer coefficient (equations 2.5 and 2.18), is proportional to U_{mf} which is itself a strong

Table 2.4 Summary of interphase transfer results for interacting bubbles (from Sit and Grace[35])

Authors	$\bar{d}_p$ (μm)	U_{mf} (cm/s)	D_G (cm²/s)	d_b (cm)	*Column*	*Overall % inc. over single bubble*	*Bubble alignment*
Sit and Grace[35]	390	11.2	0.2	12–15	2-D	34	vertical, oblique
Toei *et al.*[36]	320	8.2	0.15	6–15	2-D	35	vertical
Toei *et al.*[36]	160	2.47	0.15	5–12	2-D	15	vertical
Pereira[37]	93	0.58	0.7	4–6	3-D	3–8	oblique
Sit and Grace[34]	90	0.9	0.2	15	2-D	small	vertical

function of particle size, and so this can become the dominant term as particle size increases. Sit and Grace postulated that coalescence is accompanied by an increase in the throughflow component of transfer while the diffusive component is little affected. They suggested applying an enhancement factor of 1.8 to the throughflow term for 40 per cent of the bubbles in the bed at any instant (i.e. the approximate proportion undergoing coalescence). This leads to an overall enhancement factor of $(1.8 \times 0.4) + 0.6 = 1.32$, so that for freely bubbling beds equations 2.20b and 2.21b become for spherical bubbles:

$$\bar{k}_{GT} = \frac{U_{mf}}{3} + \left(\frac{4D_G\varepsilon_{mf}\bar{U}_b}{\pi\bar{d}_b}\right)^{1/2} \qquad (2.22)$$

and

$$\bar{K}_{GT} = \frac{2U_{mf}}{\bar{d}_b} + \frac{12}{\bar{d}_b^{3/2}}\left(\frac{D_G\varepsilon_{mf}\bar{U}_b}{\pi}\right)^{1/2} \qquad (2.23)$$

where $\bar{U}_b$ is an average bubble velocity and $\bar{d}_b$ an average bubble diameter. Equations 2.22 and 2.23 would seem to give the best currently available values of interphase mass transfer coefficients in three-dimensional beds of non-adsorbing particles. In the presence of adsorption effects however it is necessary to employ modified values according to the methods now to be outlined.

2.2.2.3 Adsorption effects

It has long been recognized that interphase mass transfer can be enhanced in cases where gas is adsorbed on the bed particles[26,38-42]. This was shown qualitatively by Cazes[43] who photographed single bubbles of adsorbing and non-adsorbing gases injected into a two-dimensional bed of porous silica–alumina cracking catalyst. *Figure 2.4* shows the much smaller bubbles that resulted from the injection of strongly adsorbing Freon-12 compared with those from an equal volume of non-adsorbing helium. No difference was found when the same gases were introduced into a bed of non-adsorbing particles of ballotini.

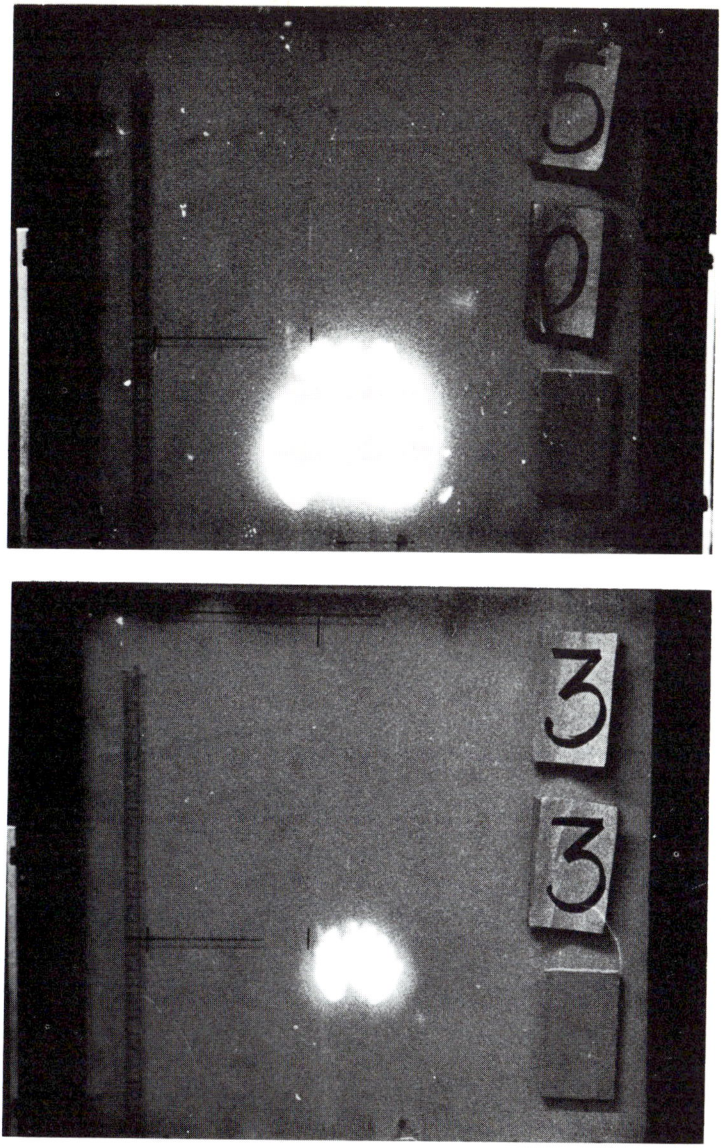

(a)

(b)

Figure 2.4 Adsorption of bubble gas by emulsion-phase solids (From Cazes[43]) (a) Helium in silica-alumina; (b) Freon-12 in silica-alumina

Chiba and Kobayashi[26] derived a theoretical expression for the ratio of interphase mass transfer coefficients in the presence (K'_{be}) and absence (K_{be}) of adsorption effects:

$$\frac{K'_{be}}{K_{be}} = \left[1 + \frac{2}{3}\, m \left(\frac{1 - \varepsilon_{mf}}{\varepsilon_{mf}}\right)\left(\frac{1}{2} + \frac{\alpha}{\alpha + 1}\right)\right]^{1/2} \tag{2.24}$$

where m, the adsorption equilibrium constant, is the ratio of the concentration of gas on the particle surface, C_{pe}, to that in the gas phase, C:

$$m = \frac{C_{pe}}{C} \tag{2.25}$$

and

$$\alpha = \frac{U_b \varepsilon_{mf}}{U_{mf}} \tag{2.26}$$

The form of equation 2.24 was confirmed experimentally by Bohle and van Swaaij[44] who measured mass transfer coefficients for a number of adsorbing (e.g. propane) and non-adsorbing (e.g. helium) gases in a fluidized bed of silica-alumina (*Figure 2.5*). The enhancement of the transfer process in the case of the former gases is apparent.

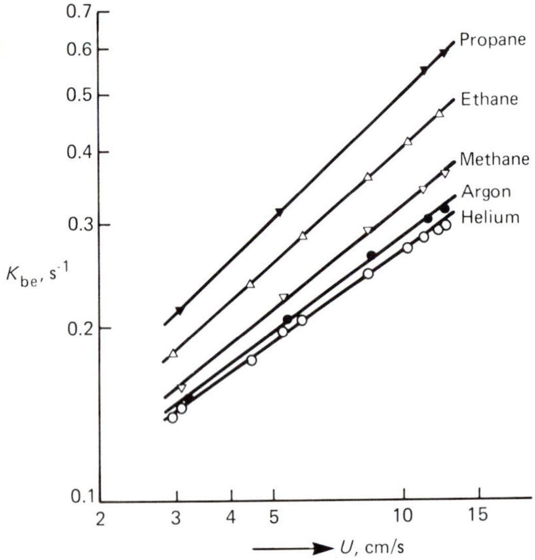

Figure 2.5 Dependence of interphase mass transfer rate K_{be} on gas adsorptivity (From Bohle and van Swaaij[44])

The importance of this effect in the formulation of fluid-bed reactor models is quite clear since adsorption-enhanced mass transfer will lead to improved gas–solid contact and hence to greater reactant conversion. There is however a compensating effect caused by mixing in the emulsion phase and this will be considered next.

2.2.3 Emulsion gas mixing

It seems reasonable to suppose that the gas flow through a bed at the point of incipient fluidization approximates closely to plug flow. In this condition Reynolds numbers based on particle diameter are very low and the gas may be pictured as percolating between the more-or-less stationary particles with little tendency to backmix. However, when at higher gas velocities, bubbles form and bed solids are moved about in their wakes, it would not be surprising if axial mixing of emulsion phase gas occurred and that, at some velocity characteristic of the particular system in question, the degree of backmixing became intense and of the same order as that of the solids themselves. Stephens *et al.*[45] were the first to investigate the backmixing caused by bubbles but later a number of other workers, notably Kunii and Levenspiel[24], Latham *et al.*[46], Potter[47] and Kato and Wen[25], also considered the problem.

The backmixing model of Kunii and Levenspiel is shown diagramatically in *Figure 2.6*. Here the upward flow of solids in cloud wakes is compensated for by a corresponding downward flow of solids in the emulsion

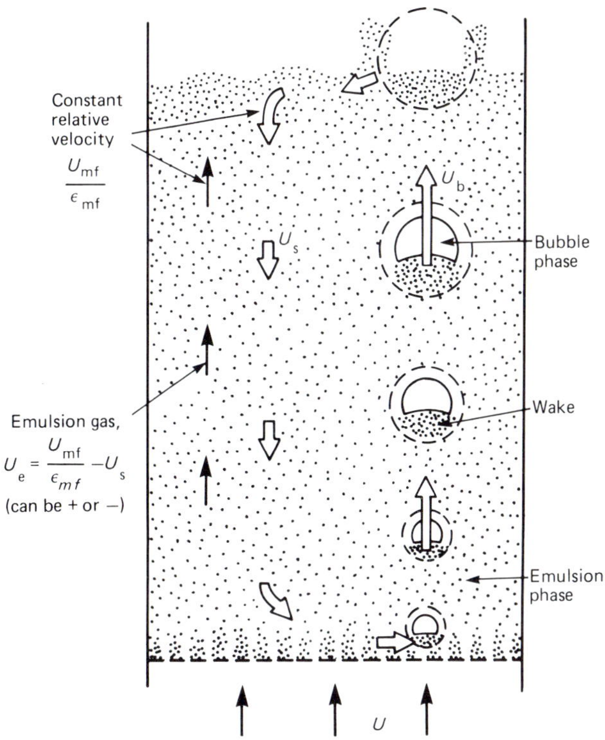

Figure 2.6 Main features of solid movement and gas flow in the bubbling bed model (From Kunii and Levenspiel[24])

phase. Assuming that the relative velocity between emulsion gas and particles remains constant at U_{mf}/ε_{mf}, the authors showed from continuity that:

$$U_e = \frac{U_{mf}}{\varepsilon_{mf}} - \left(\frac{\alpha U}{1 - f_b - \alpha f_b} - \alpha U_{mf}\right) \tag{2.27}$$

where U_e is the velocity of emulsion gas, α is the ratio of wake volume to bubble volume (V_w/V_b), and f_b is the fraction of bed occupied by bubbles. As the superficial gas flow through the bed, U, is increased, a critical velocity is reached at which the compensating downward flow of emulsion solids is equal to the upward flow of emulsion gas; at fluidizing gas velocities above this point U_e becomes negative and flow reversal leading to backmixing occurs. The critical velocity is reached when:

$$\frac{U}{U_{mf}} \geqslant (1 - f_b - \alpha f_b)\left(1 + \frac{1}{\alpha \varepsilon_{mf}}\right) \tag{2.28}$$

and assuming values of $\alpha = 0.2 - 0.4$, $\varepsilon_{mf} = 0.5$ with $f_b \ll 1$, Kunii and Levenspiel predict backmixing for fluidizing gas velocities of between six and eleven times the minimum fluidization value. Stephens *et al.*[45] obtained a similar expression to equation 2.28 for the critical backmixing velocity.

In addition to this effect of the gross circulation of solid particles, gas mixing in a fluid bed is strongly influenced by adsorption processes. This

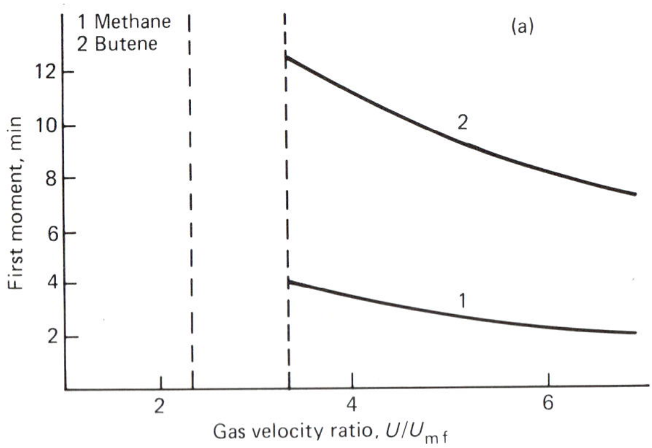

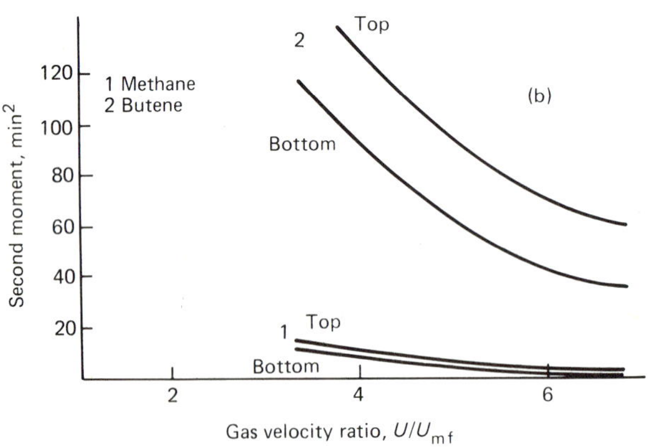

Figure 2.7 (From Yates and Constans[41])

was demonstrated by Yates and Constans[42] who injected pulses of various gases into a bed of fluidized silica–alumina and noted the increase in variance of the residence time distribution curves with increasing adsorption (*Figure 2.7*). Chiba and Kobayashi[48] incorporated the effect into a fluid-bed reactor model by assuming that the adsorbed gas was confined to particles in the bubble–cloud region and was dispersed longitudinally as part of the bubble flow. This would be expected to cause increased bypassing, and hence lower chemical conversion; the 'bypassing fraction', v_m, was given by:

$$v_m = \frac{mU_b f_w f_{bw}}{U(1 - f_{bw})} \tag{2.29}$$

where m is the adsorption partition coefficient (compare equation 2.25), f_w is the volume fraction of wake and f_{bw} is the volume fraction of bubble void and wake. However, as has already been mentioned, adsorption can increase the inter-phase gas transfer and as this will lead to greater gas–solid contact, and hence greater conversion, the adsorption-enhanced bypassing can be more-or-less counteracted. The combination of the two effects on reactor performance is thus quite complicated and depends not only on the strength of adsorption, i.e. on the value of m, but also on the operating conditions. There is a need for more experimental work in this area.

2.2.4 Division of gas flow between phases

The two-phase theory of fluidization forms the basis of many of the models described in *Table 2.2* but, as was discussed in Chapter 1 there are many systems in which the theory is demonstrably wrong and in which the emulsion-phase flow exceeds to a greater or lesser extent that predicted by the simple equations. Rowe[49] has explored some of the consequences of this for reactor modelling and has shown that in all cases where more gas flows interstitially than is allowed by the two-phase theory, chemical conversion is increased throughout the bed. This is only to be expected but at the present time there is no reliable way of calculating *a priori* how a reactant gas will divide on entry to a fluidized bed. It is known that the addition of fine particles ($\bar{d}_p < 45$ μm) to a bed of coarser material can cause an increase in emulsion-phase voidage and a corresponding rise in the interstitial gas flow[50]. This results in increased gas–solid contact, smaller bubbles and smoother fluidization and is a desirable feature in many applications. The exact result of modifying the particle size distribution in this way however cannot yet be predicted and represents another aspect of fluidization that requires more study.

2.3 Some individual models

2.3.1 Introduction

Owing to the great importance of solid-catalysed reactions in fluid-bed technology the majority of published models are designed for application to these systems although in recent years a number of studies of non-catalytic processes have been reported. Furthermore the complexities introduced by reactions which have either kinetics other than first order, have reversible or consecutive steps or which result in volume changes are commonly ignored and in general attempts are made to obtain model equations that are capable of analytical solution. A number of models are

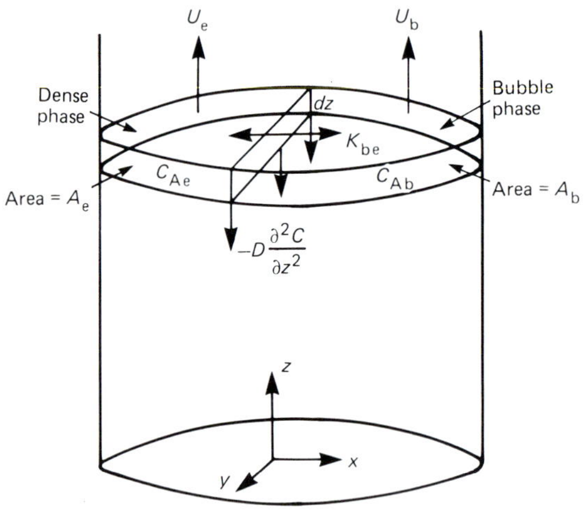

Figure 2.8 The general one-dimensional, two-phase model

based on the general one-dimensional two-phase model shown in *Figure 2.8*. For the case of an irreversible, first order gas solid-catalysed reaction with no accompanying volume change, a mass balance on reactant A moving through the emulsion phase involves five terms as follows:

$$\frac{\partial C_{Ae}}{\partial t} - E\frac{\partial^2 C_{Ae}}{\partial z^2} + U_e\frac{\partial C_{Ae}}{\partial z} + K_{be}(C_{Ae} - C_{Ab}) + k_e C_{Ae} = 0 \quad (2.30)$$

transient response − *back diffusion* + *bulk flow* + *interphase exchange* + *chemical reaction*

where C_{Ae} = concentration of A in the emulsion phase
U_e = linear velocity of emulsion gas
E = gaseous diffusion coefficient
K_{be} = interphase mass transfer rate per unit volume of bubble gas
k_e = reaction rate constant for the emulsion phase

The corresponding equation for the bubble-plus-cloud phase is:

$$\frac{\partial C_{Ab}}{\partial t} - D\frac{\partial^2 C_{Ab}}{\partial z^2} + U_b\frac{\partial C_{Ab}}{\partial z} + K_{be}(C_{Ab} - C_{Ae}) + k_b C_{Ab} = 0 \quad (2.31)$$

The equations may be simplified by making some or all of the following assumptions:

(i) the reactor operates in the steady state; this causes the transient response terms to disappear.
(ii) gas is in plug flow in both phases; hence the back diffusion terms vanish.
(iii) no reaction occurs in the bubble phase; this causes the $k_b C_{Ab}$ term to disappear from equation 2.31.

The resulting simplified relationships are simultaneous linear first-order differential equations which may be solved analytically:

$$U_e\frac{dC_{Ae}}{dz} + K_{be}(C_{Ae} - C_{Ab}) + k_e C_{Ae} = 0 \quad (2.32)$$

$$U_b\frac{dC_{Ab}}{dz} + K_{be}(C_{Ab} - C_{Ae}) = 0 \quad (2.33)$$

2.3.2 The Davidson–Harrison model[2]

This was one of the earliest models to be based on bubble dynamics and it rests on the following assumptions:

(i) emulsion-phase gas flows with a superficial velocity U_{mf} and is either completely mixed or in plug flow;
(ii) bubbles are of uniform size throughout the bed;
(iii) reaction takes place only in the emulsion with first-order kinetics and a rate constant k (s^{-1});
(iv) interphase gas transfer occurs by a combined process of molecular diffusion and throughflow, the overall exchange rate (volume of gas per unit volume of bubble per unit time) being given by equations 2.7 and 2.8.

The model was developed before the importance of bubble clouds was realized and it makes no reference to them.

The model equations are developed for the two phases separately by carrying out a mass balance on the reactant in each phase. Thus for the case of the completely mixed emulsion phase the concentration of reactant leaving this phase is given by:

$$C_{Ae} = \frac{C_{Ao}(1 - f_b e^{-X})}{1 - f_b e^{-X} + k'} \quad (2.34)$$

where X, the interphase exchange parameter is:

$$X = \frac{K_{be}H}{U_b} \tag{2.35}$$

H being the bed height and U_b the bubble velocity.
Also:

C_{Ao} = inlet reactant concentration,

$$f_b = 1 - \frac{U_{mf}}{U} \tag{2.36}$$

and

$$k' = \frac{kH_{mf}}{U} \tag{2.37}$$

For the bubble phase:

$$C_{Ab} = C_{Ae} + (C_{Ao} - C_{Ae})e^{-X} \tag{2.38}$$

The concentration of unconverted reactant, C_{AH}, leaving the bed as a whole is then found from:

$$C_{AH} = f_b C_{Ab} + (1 - f_b)C_{Ae} \tag{2.39}$$

and substituting C_{Ab} and C_{Ae} in equation 2.39, the overall fraction unconverted becomes:

$$\frac{C_{AH}}{C_{Ao}} = f_b e^{-X} + \frac{(1 - f_b e^{-X})^2}{1 - f_b e^{-X} + k'} \tag{2.40}$$

From equation 2.40 it is apparent that as $X \rightarrow \infty$, the performance of the reactor approaches that of a completely stirred tank reactor (CSTR), i.e.

$$\frac{C_{AH}}{C_{Ao}} \rightarrow \frac{1}{1 + k'} \tag{2.41}$$

which is consistent with the assumption of a completely mixed emulsion phase.
Furthermore, for large values of k:

$$\frac{C_{AH}}{C_{Ao}} \rightarrow f_b e^{-X} \tag{2.42}$$

i.e. even for a very fast reaction, owing to the bypassing effect of bubbles, there will always be some reactant leaving the bed unreacted. Some typical predictions of reactant concentration profiles are shown in *Figure 2.1* and have already been referred to.

The second form of the model in which the emulsion phase gas is assumed to be in plug flow leads to the somewhat more complicated expression:

$$\frac{C_{AH}}{C_{Ao}} = \frac{1}{m_1 - m_2}\left[m_1 e^{-m_2 H}\left(1 - \frac{m_2 H U_{mf}}{XU}\right) - m_2 e^{-m_1 H}\left(1 - \frac{m_1 H U_{mf}}{XU}\right)\right] \tag{2.43}$$

where m_1 and m_2 are obtained from:

$$2H(1-f_b)m = (X + k') \pm [(X + k')^2 - 4k'X(1-f_b)]^{1/2} \tag{2.44}$$

and where $m = m_1$ with the positive sign and $m = m_2$ with the negative sign.

The expressions for the fraction of reactant unconverted from both forms of the model are functions of three main variables f_b, k' and X of which the latter is the most difficult to estimate. A value for X may however be obtained by use of the Davidson–Harrison interphase gas exchange equations 2.7 and 2.8 giving:

$$X = \frac{H}{0.711(gd_b)^{1/2}}\left[4.5\left(\frac{U_{mf}}{d_b}\right) + 5.85\left(\frac{D_G{}^{1/2}g^{1/4}}{d_b{}^{5/4}}\right)\right] \tag{2.45}$$

which, provided values of the equivalent bubble diameter, d_b, are known, enables X to be found.

Although the authors point out that the theory used for calculating the gas exchange parameter does not have strong experimental support (*see* Section 2.2.2) the model predictions have been found to agree quite well with some published experimental data[2]. The model has been extended by Darton[51] to take account of subsequent developments in the theory of bubble growth in beds of fine particles.

2.3.3 Partridge–Rowe model[21]

The differs fundamentally from the Davidson–Harrison model in that it takes account of gas–solid contact and reaction in the cloud region surrounding a bubble. It assumes the emulsion-phase gas to be in plug flow and to react according to:

$$\frac{-dC_A}{dt} = k_e C_A^n \tag{2.46}$$

where

$$k_e = k\varepsilon_{mf};$$

k is a rate constant determined from fixed-bed experiments and n is the order of reaction. Bubbles are assumed to be of the form described by Murray[32,33] but modified (as shown in *Figure 2.9*) to take account of the

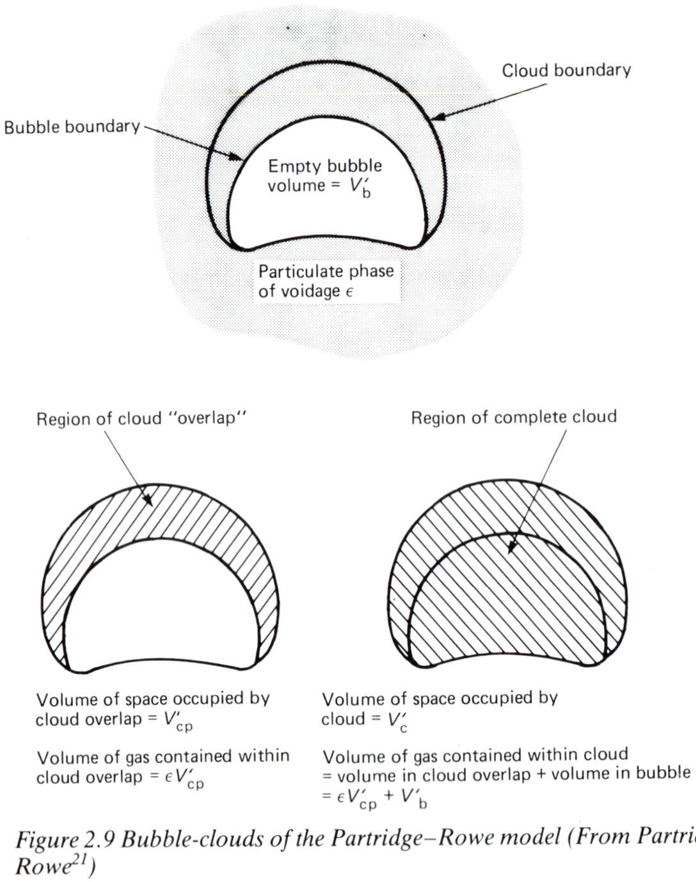

Figure 2.9 Bubble-clouds of the Partridge–Rowe model (From Partridge and Rowe[21])

wake which is included in the emulsion phase. From X-ray observations of bubbles in three-dimensional beds Rowe and Partridge[52] concluded that the fraction of the bubble sphere occupied by wake, f_w, was 0.25. Gas in the bubble-plus-cloud region is then assumed to be perfectly mixed and to react according to equation 2.46 but with a rate constant that allows for the extent of gas–solid contact given by:

$$k_c = \frac{k\varepsilon_{mf}V'_{cp}}{V'_{cp} + V'_b} \tag{2.47}$$

An empirical relationship linking the volume of the cloud overlap region with the ratio of bubble velocity to emulsion-phase gas velocity, α, was found to be:

$$\frac{V'_{cp}}{V'_b} = \frac{1.17}{\alpha - 1} \tag{2.48}$$

and thus the cloud-phase rate constant becomes:

$$k_c = \frac{1.17k\varepsilon_{mf}}{0.17 + \alpha} \tag{2.49}$$

Variation of bubble diameter with bed height is allowed for via empirically determined variations in the value of α with height (*see* below).

Mass balance equations for the two phases then follow directly from equations 2.30 and 2.31:

$$U_e \frac{dC_{Ae}}{dz} + K_{bce} \frac{A_c}{A_e} (C_{Ae} - C_{Ac}) + k_e C_{Ae} = 0 \tag{2.50}$$

$$U_c \frac{dC_{Ac}}{dz} + K_{bce}(C_{Ac} - C_{Ae}) + k_c C_{Ac} = 0 \tag{2.51}$$

For non-first-order kinetics the equations must be solved by numerical methods[53] but for a first-order reaction they may be reduced to a linear second-order differential equation in C_{Ac}:

$$\frac{d^2C_{Ac}}{dz^2} + \frac{JdC_{Ac}}{dz} + KC_{Ac} = 0 \tag{2.52}$$

$$\text{where } J = \frac{(K_{bce} + k_c)}{U_c} + \frac{K_{bce}A_c}{A_eU_e} + \frac{k_e}{U_e} \tag{2.53}$$

$$\text{and } K = \frac{k_cK_{bce}A_c}{U_cU_eA_e} + \frac{k_e}{U_cU_e}(K_{bce} + k_c) \tag{2.54}$$

The auxiliary equation is:

$$S^2 + JS + K = 0 \tag{2.55}$$

with roots S_1 and S_2:

$$S_1 = 0.5[-J + (J^2 - 4K)^{1/2}]$$
$$S_2 = 0.5[-J - (J^2 - 4K)^{1/2}] \tag{2.56}$$

The solution to equation 2.52 then becomes:

$$C_{Ac} = Z_1 \exp(S_1 z) + Z_2 \exp(S_2 z) \tag{2.57}$$

in which the values of Z_1 and Z_2 may be found from the boundary conditions.

The final expressions for the unconverted fraction of reactant emerging from each phase are:

$$\frac{C_{Ac}}{C_{Ao}} = \frac{1}{U_c(S_1 - S_2)}\left[(k_c + S_1 U_c)\exp(S_2 z) - (k_c + S_2 U_c)\exp(S_1 z)\right] \tag{2.58}$$

$$\frac{C_{Ae}}{C_{Ao}} = \frac{1}{K_{bce}}\left[\frac{S_2}{S_1 - S_2}(k_c + S_1 U_c)\exp(S_2 z) - \frac{S_1}{S_1 - S_2}(k_c + S_2 U_c)\exp(S_1 z) + (K_{bce} + k_c)\right] \tag{2.59}$$

The total outlet conversion is then found by averaging these two fractions in proportion to the division of flow.

A vital parameter in the model is $\bar{\alpha}_F$, the mean value weighted on a cloud–gas volume basis of the ratio of bubble velocity to interstitial gas velocity. This is used to calculate the area of bed cross-section occupied by bubble clouds:

$$A_c = \frac{Q_b \varepsilon_{mf}}{(\bar{\alpha}_F - 1)U_{mf}} \tag{2.60}$$

where the bubble flow, Q_b, is found from the two-phase theory:

$$Q_b = A(U_z - U_{mf}) \tag{2.61}$$

and

$$U_z = U_{top}\left[\frac{1}{1 + \left(1 - \frac{z}{H}\right)\frac{\Delta p}{P_{top}}}\right] \tag{2.62}$$

Here U_{top} and p_{top} are respectively the superficial gas velocity and pressure at the top of the bed, and Δp is the pressure drop across the bed. The emulsion-phase area is then:

$$A_e = A - A_c \tag{2.63}$$

The values of $\bar{\alpha}_F$ used in the model calculations were obtained experimentally for various powders by cine photography of a freely-bubbling two-dimensional bed. Extrapolation of these values to a three-dimensional system however is a source of weakness in the model since it appears to cause an overestimate of cloud gas flow (*see* below).

Interphase gas exchange in this model is estimated using the analogy with transfer between a drop of one imiscible liquid and another through which it is rising (Section 2.2.2.1) and which leads to equation 2.14 for the exchange parameter K_{bce}.

An experimental test of the model was carried out by Ellis *et al.*[54] using the dehydrogenation of butenes in a 0.10 m diameter reactor. They concluded that the yields predicted by the model were in reasonable agreement with experimental observations although it was found necessary to adjust arbitrarily the values of $\bar{\alpha}_F$ at low bed levels since if this were not done the calculated flow of cloud gas more than filled the bed cross-section. A similar problem was encountered with this model by Chavarie and Grace (*see* Section 2.4).

2.3.4 The Kunii-Levenspiel bubbling bed model[24]

This model can be cast in several forms depending on the type of gas–solid reaction under consideration, i.e. with porous or non-porous catalyst, with or without mass transfer limitations, etc. Its principal assumptions however, are:

(i) reactant gas it is transported through the bed in uniformly sized bubbles with associated clouds and wakes; gas flow through the emulsion is negligably small.
(ii) bubble gas is in plug flow.
(iii) interphase gas exchange occurs in two stages, from bubble to cloud and from cloud to emulsion. The corresponding exchange rates are given by equations 2.8 and 2.15, in Section 2.2.2.1.

The fraction of reactant emerging unconverted is then simply:

$$\frac{C_A}{C_{Ao}} = \exp(-k_f) \tag{2.64}$$

where

$$k_f = \frac{Hk_r}{U_b}\left[\gamma_b + \cfrac{1}{\cfrac{k_r}{K_{bc}} + \cfrac{1}{\gamma_c + \cfrac{1}{\cfrac{k_r}{K_{ce}} + \cfrac{1}{\gamma_e}}}}\right] \tag{2.65}$$

If the reaction is a first order catalytic process with no mass transfer limitation then, based on the rate of change of N_A, the number of moles of reactant A, per unit volume of porous solid V_s:

$$-\frac{1}{V_s}\frac{dN_A}{dt} = k_r C_A \quad (2.66)$$

The terms γ_b, γ_c and γ_e in equation 2.65 are the dimensionless ratios of solids dispersed in bubbles, clouds and emulsion respectively to the total volume of bubbles in the bed:

$$f_b(\gamma_b + \gamma_c + \gamma_e) = (1 - \varepsilon_{mf})(1 - f_b) \quad (2.67)$$

where f_b is the fraction of bed occupied by bubbles. In beds containing large, fast bubbles:

$$f_b \simeq \frac{U - U_{mf}}{U_b} \quad (2.68)$$

where

$$U_b = 0.711(gd_b)^{1/2} + U - U_{mf} \quad (2.69)$$

For most practical purposes, γ_b may be taken as being negligibly small while γ_c is found from:

$$\gamma_c = (1 - \varepsilon_{mf})\left[\frac{3U_{mf}/\varepsilon_{mf}}{0.711(gd_b)^{1/2} - U_{mf}/\varepsilon_{mf}} + \frac{V_w}{V_b}\right] \quad (2.70)$$

V_b and V_w are the volumes of bubble and wake respectively given as a function of bed particle diameter by the empirical correlation of Rowe and Partridge[53] shown in *Figure 2.10*. The value of γ_e is found from equation 2.67 by substitution of γ_c.

The model is simple and straightforward to use although it does, of course, contain many simplifying assumptions. One in particular that deserves mention is the assumption that the contribution of emulsion phase reaction to the overall performance may be ignored. Kunii and Levenspiel maintain that at gas velocities in excess of $3U_{mf}$ the fraction of gas flowing interstitially is very small and so the overall conversion may be approximated by that of the bubble phase alone. This has been questioned by a number of workers, and in particular by Potter *et al.* who have developed a closely similar model but one which treats the system in a more rigorous way (Section 2.3.5).

Experimental test of the bubbling bed model is discussed in Section 2.4.

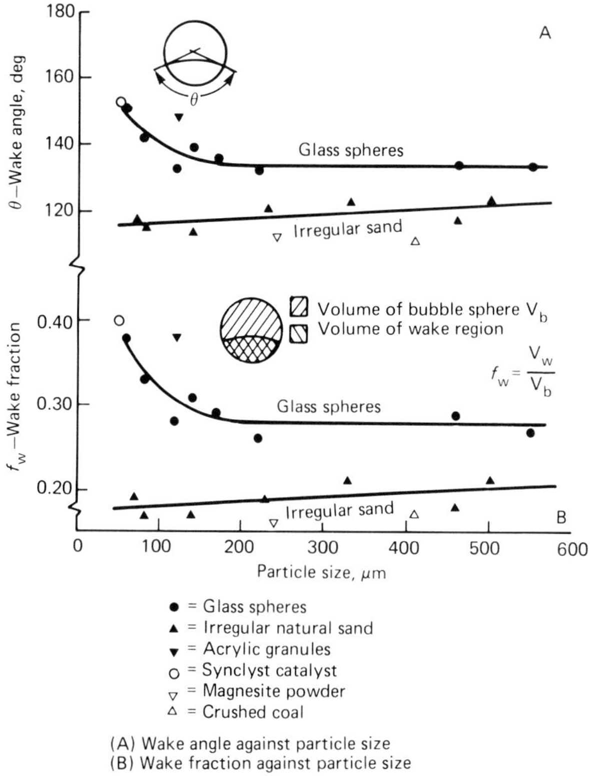

Figure 2.10 Variation of wake angle and wake fraction with particle size (From Rowe and Partridge[52])

2.3.5 The countercurrent backmixing model

This model has been developed by Potter and his co-workers and is essentially an elaborated version of the foregoing bubbling bed model. Stephens, Sinclair and Potter[45] showed that at gas velocities in excess of a critical value U_{cr}, where:

$$\frac{U_{cr}}{U_{mf}} = 1 + \frac{1}{\varepsilon_{mf} f_w} \qquad (2.71)$$

strong backmixing of emulsion-phase gas occurs as a result of the downward flow of solids compensating for the upward flow of material in bubble wakes (Section 2.2.3). Potter subsequently introduced an additional term to take account of the space occupied by bubbles and wakes[55]:

$$\frac{U_{cr}}{U_{mf}} = 1 + \frac{1}{\varepsilon_{mf} f_w} [1 - f_b(1 + f_w)] \qquad (2.72)$$

The emulsion phase flow U_e is then negative when U exceeds U_{cr}, and overall:

$$U = U_b + U_c + U_e \tag{2.73}$$

where the gas in the cloud wake is considered to be a separate phase moving with velocity U_c. It is assumed that fresh gas enters the bed as bubbles and that at $U > U_{cr}$ the wake gas is a mixture of fresh and backmixed gas. Gas leaving the bed surface is considered to be made up of all the bubble gas and a proportion of the cloud-wake gas, the remainder of which provides the down-flowing gas of the emulsion.

Material balance equations are set up for reactant A in the three phases:

bubble phase

$$-\frac{dC_{Ab}}{dz} = \frac{K_{bc}(C_{Ab} - C_{Ac})f_b}{U_b} \tag{2.74}$$

cloud-wake phase

$$-\frac{dC_{Ac}}{dz} = \frac{K_{ce}(C_{Ac} - C_{Ae})f_b + K_{bc}(C_{Ac} - C_{Ab}) + kC_{Ac}f_w f_b}{U_c} \tag{2.75}$$

emulsion phase

$$-\frac{dC_{Ae}}{dz} = \frac{K_{ce}(C_{Ae} - C_{Ac})f_b + kC_{Ae}[1 - f_b(1 + f_w)]}{U_e} \tag{2.76}$$

In these equations:

(i) the interphase exchange parameters are, as in the case of the Kunii-Levenspiel model, identical with those given in Section 2.2.2.1.
(ii) bubble rise velocity is given by equation 2.69
(iii) bubble diameters are assumed to increase linearly with bed height according to

$$d_b = d_o + mz \tag{2.77}$$

where, on the basis of experimental evidence[27]:

$$d_o = 1.08(U - U_{mf})^{0.33} \text{ (cgs units)} \tag{2.78}$$

and

$$m = 2.05 \times 10^{-2}\, U \text{ (cgs units)} \tag{2.79}$$

(iv) following Woolard and Potter[56]:

$$f_w = 0.476\exp(-168.5d_p) \tag{2.80}$$

where the particle diameter d_p is in inches.

The concentration of reactant gas on a time averaged basis is given in terms of the fractions of phases present:

$$C_{A,av} = f_b C_{Ab} + f_b f_w C_{Ac} + [1 - f_b(1 + f_w)]C_{Ae} \tag{2.81}$$

A number of experimental tests of the model have been reported by Potter *et al.*[27,55], some of which are summarized in *Figure 2.11*. It is

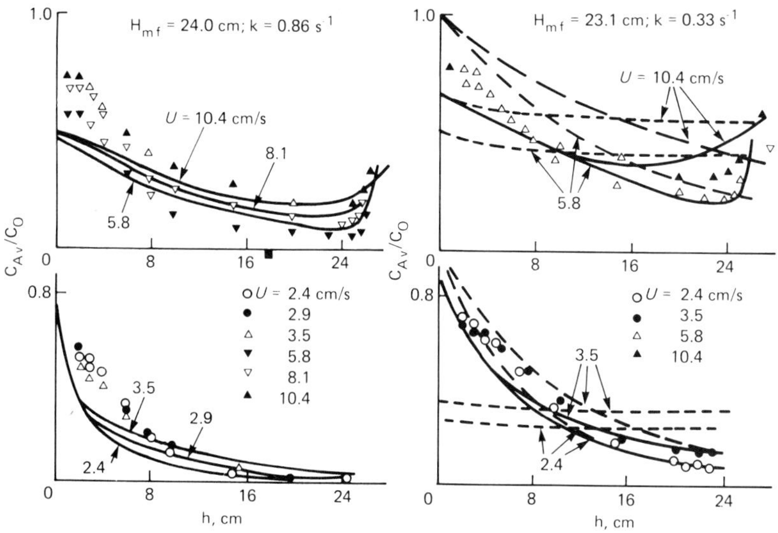

Figure 2.11 Experimental axial profiles of average concentration of ozone with bed height. From a study of the decomposition of ozone in a 22.9 cm diameter reactor (Fryer and Potter[27]). The lines show the predictions of the counter current backmixing model

noteworthy that the reactant concentration decreases monotonically with height when $U < U_{cr}$, but when $U > U_{cr}$ the curve passes through a minimum value at a point below the bed surface. In all cases agreement is good. Further support for the model has come from recent studies of o-xylene oxidation in a slugging bed reactor[57].

2.3.6 The Kato–Wen bubble assemblage model

As with the Partridge–Rowe model, variation of bubble size with height above the distributor is allowed for, although in the Kato–Wen model the equation for this variation is based on the empirical correlation of Kobayashi *et al.*[58]:

$$d_b = 1.4 \rho_p d_p \left(\frac{U}{U_{mf}} \right) z \tag{2.82}$$

with an additional term for the initial bubble diameter at the distributor derived from the work of Cooke *et al.*[59]:

$$d_{bo} = \left[\frac{6(U - U_{mf})}{\pi n_0} \right]^{0.4} g^{-0.2} \tag{2.83}$$

Here n_0 is the number of holes per unit surface area of distributor; for a porous plate distributor n_0 is taken to be so large that $d_{bo} \simeq 0$. The model then assumes that the bed can be divided into a number of vertical compartments, the height of each being equal to the bubble diameter at the corresponding bed height. Bubble and emulsion phase gas is assumed to be completely mixed within each phase, the interphase transfer coefficient per unit volume of bubble, K_o, being given by the relationship of Kobayashi *et al.*[60]:

$$K_o = \frac{11}{d_b} \quad \text{(cgs units)} \tag{2.84}$$

Gas bubbles are assumed to be surrounded by clouds with radii given by the equation of Davidson[2]:

$$\left(\frac{R_c}{R_b} \right)^3 = \frac{U_b + 2U_{mf}/\varepsilon_{mf}}{U_b - U_{mf}/\varepsilon_{mf}} \tag{2.85}$$

where

$$U_b = 0.711(gd_b)^{1/2} \tag{2.86}$$

(*See* Chapter 1, Section 1.6.2)

Material balances on the reactant in the bubble and emulsion phases of the nth compartment are then as follows:

bubble phase

$$(AUC_{Ab})_{n-1} = [K_0'V_b(C_{Ab} - C_{Ae})]_n + (r_bV_c)_n + (AUC_{Ab})_n \tag{2.87}$$

emulsion phase

$$[K_o'V_b(C_{Ab} - C_{Ae})]_n = (r_eV_e)_n \tag{2.88}$$

where r_b and r_e are reaction rates per unit volume in bubble and emulsion phases respectively; V_c, V_b and V_e are the respective volumes of cloud, bubble and emulsion, and K_o' is the interchange coefficient based on unit volume of bubble-cloud space:

$$K_o' = \frac{K_o(U_b - U_{mf}/\varepsilon_{mf})}{U_b + 2U_{mf}/\varepsilon_{mf}} \tag{2.89}$$

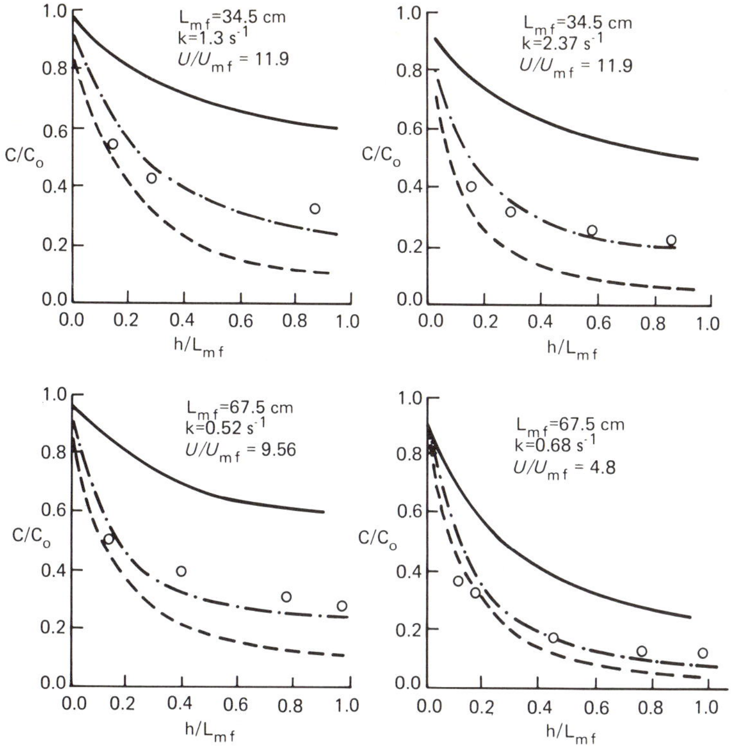

Figure 2.12 Comparison of experimental ozone concentration profile with calculated values. Bubble-phase concentration (———); Average concentration (- - - - -); Emulsion-phase concentration (– – – –); Experimental data (○)[61] (From Kato and Wen[25])

Model predictions were found to agree reasonably well with a number of published experimental results. One such comparison is shown in *Figure 2.12* where results obtained by Tanaka[61] on the catalytic decomposition of ozone are plotted with the model predictions of ozone concentration profiles in both bubble and emulsion phases; also plotted is an average concentration found from:

$$C_{A,av} = \frac{H - H_{mf}}{H_{mf}} C_{Ab} + \frac{H_{mf}}{H} C_{Ae} \tag{2.90}$$

The model has the advantages of containing no adjustable parameters and of being of a form suitable for computer simulation.

2.3.7 The Werther model[62]

The basic assumption of this model, and what sets it apart from the others so far considered, is that interphase mass transfer may be described in the same way as is done for gas–liquid systems, i.e. as an adsorption of reactants from the bubble phase followed by a pseudo-homogeneous reaction in the emulsion. The transfer process can then be described in terms of film theory[63,64], in which mass transfer occurs through a film of thickness δ, where:

$$\delta = \frac{D_G}{k_g} \tag{2.91}$$

D_G being the gas diffusion coefficient (m^2/s) and k_g the mass transfer coefficient (m/s). A number of simplifying assumptions are made:

(i) emulsion-phase flow may be neglected
(ii) bubble-phase gas is in plug flow
(iii) mass transfer characteristics are independent of bed height and radial distance.

The predicted conversion of reactant is given by:

$$X_A = \left(1 - \frac{C_A}{C_{Ao}}\right) = 1 - \exp\left\{-\left[\frac{(\Phi^{-1} - 1)Ha + \tanh Ha}{(\Phi^{-1} - 1)Ha \tanh Ha + 1}\right] HaN\alpha\right\} \tag{2.92}$$

the dimensionless numbers of which are:
(a) the Hatta number, Ha :

$$Ha = \frac{\sqrt{(kD_G)}}{k_g} \tag{2.93}$$

where k is a reaction rate constant per unit volume of emulsion phase;
(b) the number of mass transfer units, N_α:

$$N_\alpha = \frac{k_g aH}{U - U_{mf}} \tag{2.94}$$

where a is the interfacial area per unit volume of bed and H is the expanded bed height;
(c) the ratio of film volume to total volume of emulsion phase, Φ:

$$\Phi = \frac{a\delta}{1 - f_b} \tag{2.95}$$

where f_b is, as before, the bubble hold up.

For the special case of $Ha \ll 1$, equation 2.92 reduces to:

$$X_A = 1 - \exp[-(N_r^{-1} + N_\alpha^{-1})^{-1}] \tag{2.96}$$

where N_r is the number of reaction units:

$$N_r = \frac{k(1 - f_b)H}{U - U_{mf}} \tag{2.97}$$

From experiments conducted with fluid beds of up to 3 m in diameter and 10 m in height, in which a capacitance probe was used to measure bubble characteristics[65,66], Werther produced empirical correlations for N_α

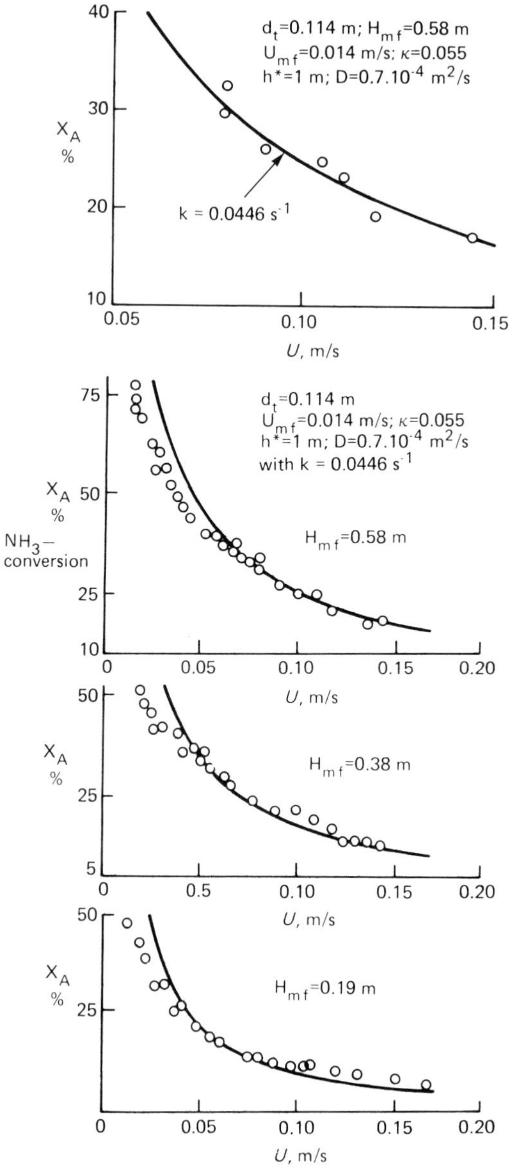

Figure 2.13 Comparison between Werther model[62] predictions (———) and experimental results (○) of Massimilla and Johnstone[67]

in terms of the height of a transfer unit, H_α, and applied the resulting model equations to be published results of other workers on the catalytic oxidation of ammonia[67], the hydrogenation of ethylene[9,68] and the catalytic decomposition of nitrous oxide[7]. In all cases good agreement was found with the model predictions, a typical example being shown in *Figure 2.13*.

2.3.8 The Shell model

Because of the problems of accurately predicting fluid-bed hydrodynamics from first principles at the present time, empirically based methods are normally used in industrial design. One approach described in some detail in the literature[69–71] is that of the Shell company whose semi-empirical model was used in the design of a catalytic reactor for the air oxidation of hydrogen chloride to chlorine. The model is based on that of van Deemter[17] which assumes:

(i) a solid-free bubble phase in plug flow
(ii) negligible gas flow through the emulsion phase
(iii) emulsion gas mixing by eddy diffusion.

Differential equations for the mass balances on the reactant in the two phases are:

bubble phase

$$\frac{\partial C_b}{\partial \xi} + N_\alpha(C_b - C_e) = 0 \qquad (2.98)$$

emulsion phase

$$N_\alpha(C_e - C_b) - \frac{1}{N_E}\frac{\partial^2 C_b}{\partial \xi^2} + N_r C_e = 0 \qquad (2.99)$$

Here ξ is the dimensionless height through the bed while C_b and C_e are reactant concentrations in bubble and emulsion respectively. The model is expressed mainly in terms of three dimensionless parameters: N_α, the number of interphase transfer units, N_E, the number of mixing units in the emulsion phase, and N_r, the number of reaction units.

$$N_\alpha = \frac{k_g' H}{U} \quad ; \quad N_E = \frac{UH}{f_e E} \quad ; \quad N_r = \frac{k' H}{U} \qquad (2.100)$$

In these expressions $k_g'(s^{-1})$ is a mass transfer coefficient, $H(m)$ is the height of the expanded bed, f_e is the volume fraction of the reactor occupied by emulsion gas, E (m^2/s) is a mixing coefficient, and $k'(s^{-1})$ is the average reaction rate constant in the expanded bed which is obtained from

$k(s^{-1})$, the rate constant per unit weight of catalyst and f_{ep} the fraction of bed occupied by emulsion phase:

$$k' = f_{ep}\, k \tag{2.101}$$

To obtain the values of these parameters for the Shell Chlorine Process an extensive programme of experiments was carried out using the process catalyst carrier, a broad range microspheroidal powder with 10 to 20 per cent fines under 44 μm. The height of a transfer unit, H_α, was found from the residence time distribution of a tracer gas injected into a series of beds of increasing size and the following correlation, expected to hold for beds of the same powder with heights and diameters of up to 10 m, was obtained:

$$H_\alpha = \left(1.8 - \frac{1.06}{D_T^{1/4}}\right)\left(3.5 - \frac{2.5}{H^{1/4}}\right) \tag{2.102}$$

where D_T is the bed diameter and H the expanded bed height, both in metres.
Then:

$$N_\alpha = \frac{H}{H_\alpha} \tag{2.103}$$

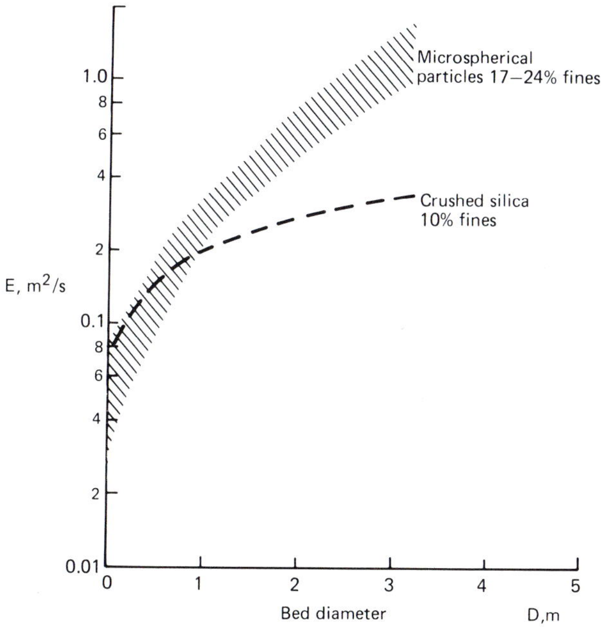

Figure 2.14 Mixing coefficients as a function of bed diameter, at 0.2 m/s gas velocity (From van Swaaij and Zuiderweg[71])

The number of diffusion units was also determined empirically using labelled catalyst particles (*Figure 2.14*). The following iterative procedure was then used to size the reactor:

> Assuming a realistic gas velocity, U, the bed diameter was calculated from the required throughput (as in the example at the beginning of the Chapter). Then a first estimate of the expanded bed height, H, was put into equation 2.102 and the corresponding value of H_α obtained. From H and the known expansion characteristics of the catalyst, the height of the unexpanded bed, and hence the amount of catalyst in the bed, was found. Knowing the catalyst activity per unit weight from fixed-bed experiments, the average reaction rate constant for the expanded bed, k', was calculated and hence the number of reaction units, N_r (equation 2.100), was obtained. Finally a value of N_E corresponding to the chosen bed diameter was found from the empirical correlations (*Figure 2.14*). The predicted conversion for this set of parameters was then found from the solutions to equations 2.98 and 2.99. Depending on the difference between this predicted value and the value required from the initial design criteria a new expanded bed height was selected and the calculation repeated.

The problem with this method is clearly the cost involved in carrying out the experimental programme to establish the necessary correlations, although once these have been established the design of a similar reactor using a similar catalyst powder is fairly straightforward. The Shell model has nevertheless recently been put on a more deterministic basis by incorporating physical interphase exchange relationships and a bubble growth mechanism[72].

2.4 Model comparisons

Few comparative studies have been reported, most experimental work having concentrated on the exploration of one particular model. Chavarie and Grace however carried out an elegant study of the catalytic decomposition of ozone in a two-dimensional bed with transparent sides, in which not only overall conversions but also reactant concentrations in the two phases could be measured[73]. Ozone concentrations in the bubble phase were determined from the extent of absorption of UV light by the bubbles which were cylindrical in shape and which filled the space between the two parallel faces of the bed. A UV source and measuring spectrophotometer were situated on opposite sides of the 2-D bed and were activated when a bubble passed between them. Reactant concentrations in the emulsion phase were measured by withdrawing gas from a number of sampling points situated along the bed axis. The visible bubble flowrate, Q_b, in the catalyst powder was determined for the range of velocities employed by cine photography of the bed and it was found that although the estimate of

Q_b given by the two-phase theory seemed to be approached at positions well above the distributor, over most of the bed it was considerably lower than expected. This result was the cause of some difficulty since all the models tested were based on the assumptions of the two-phase theory, and since the theory was inapplicable to the system it was unlikely that the model predictions would correspond closely to the observations. Thus although in principle the analytical technique provided a stringent test of the models, in practice it was severely limited from the outset.

Predictions of four groups of models–those of Davidson and Harrison, Partridge and Rowe, Kato and Wen, and Kunii and Levenspiel–were tested and the conclusions in each case were as follows:

(i) *Davidson–Harrison model.* Both versions of this model were found to show poor agreement with concentration profiles and overall conversions. The interphase transfer rate was shown to be too high and the model in general was found to be inappropriate over the range of experimental conditions investigated (*Figure 2.15 (a)*).

(ii) *Partridge–Rowe model.* The inapplicability of the two-phase theory to the reactor system produced a serious problem in this case since the calculated values of bubble-cloud area exceeded the cross-sectional area of the bed, particularly at the lower bed levels. An attempt was made to correct for this incompatability by varying the allotment of gas flow between the bubble-cloud and emulsion phases but even then, close agreement was only found in the upper reaches of the bed.

(iii) *Kato–Wen model.* Although this gave good predictions of bubble-phase profiles the agreement with those of the emulsion and with the overall conversion was poor (*Figure 2.15(b)*).

(iv) *Kunii–Levenspiel model.* This showed the best overall agreement with the data (*Figure 2.15(c)*). The authors attributed this to the model's good predictions of dense-phase concentrations which were in turn ascribed to (a) its neglect of interstitial flow, and (b) its modest rate of interphase transfer. A further factor, of a more fortuitous nature, was the use by Kunii and Levenspiel of an expression for the fraction of bed occupied by bubbles:

$$f_b = \frac{U - U_{mf}}{U_b} \tag{2.104}$$

which was found to overestimate f_b by between 28 and 44 per cent in the 2-D bed used, but which provided a convenient compensation for the undetermined amount of gas throughflow in the bubble phase; this led indirectly to a decrease in the rate of depletion of reactant in the bubbles.

Chavarie and Grace also observed that simply on the basis of its predictions of overall conversion the single-phase plug flow relationship:

$$\frac{C_A}{C_{A0}} = \exp -\left(\frac{kH_{mf}}{U}\right) \tag{2.105}$$

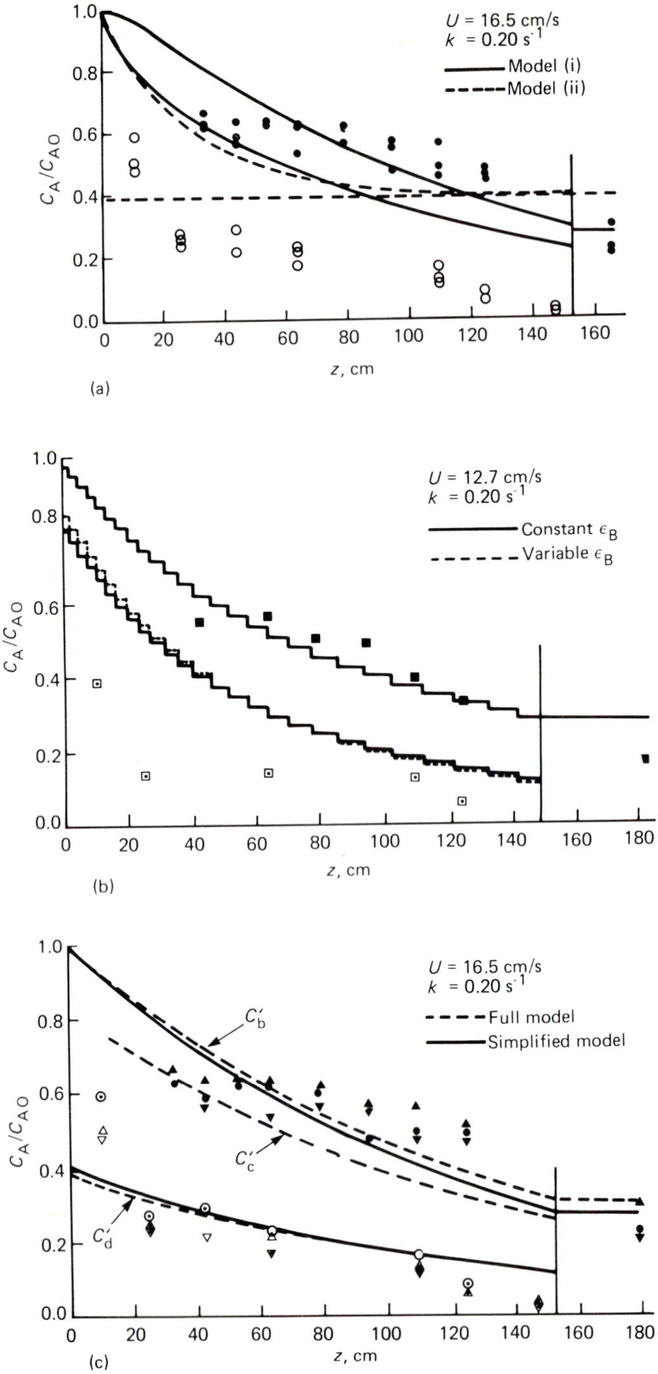

Figure 2.15 Concentration profiles and model predictions of ozone decomposition in 2-D bed (From Chavarie and Grace[73])

gave as good agreement with experimental observation as the Kunii–Levenspiel model. They added the comment however that it is only the realistic two-phase models that are likely to prove reliable over a wide range of operating conditions. A general observation made by the authors was that none of the models tested correctly accounted for the considerable end effects observed in the reactor at both the inlet and outlet (*Figure 2.15*). In other words insufficient allowance was made for reaction near the distributor and in the freeboard above the bed surface (*see* Section 2.5).

Another comparative study was carried out by Shaw *et al.*[74] who used a conventional laboratory-scale fluid-bed reactor of 0.20 m diameter to study the hydrogenolysis of butane over a nickel-on-silica catalyst. Their results showed good agreement with most model predictions although the interphase exchange parameter was adjusted empirically.

Many of the two-phase models discussed in this Chapter can, with suitable modifications, be used to model slugging fluidized bed reactors, and recently a comparison was made of three such modifications applied to the catalytic oxidation of o-xylene in a 0.10 m diameter bed[57]. The three

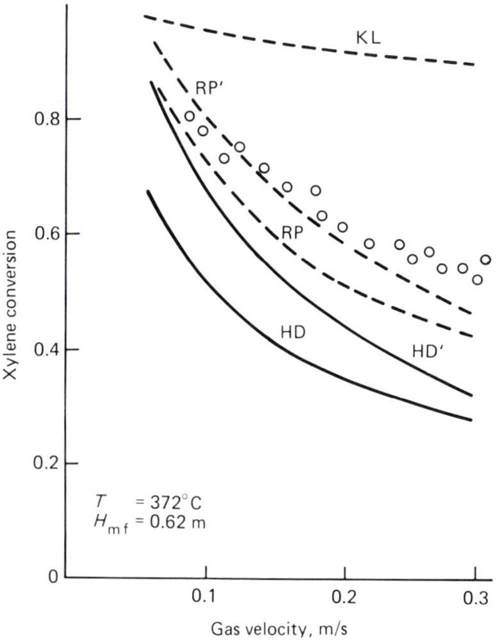

Figure 2.16 Comparison between model predictions and experiment, for o-xylene oxidation in a slugging bed reactor (From Yates and Gregoire[57]) Key: KL, Kunii–Levenspiel; HD, Hovmand–Davidson; RP, Raghuraman–Potter; HD', modified HD; RP', modified RP; (○), experimental points

models considered were the Davidson–Harrison as modified by Hovmand and Davidson[75], the Raghuraman–Potter version of the countercurrent backmixing model[76], and the Kunii–Levenspiel model. Some results of the study are shown in *Figure 2.16* from which it is clear that the Raghuraman–Potter model most closely fits the experimental points, although in every case it was found necessary to add a term to account for reaction in the lower regions of the bed where slugs had not yet developed.

There is clearly a need for more comparative studies in which one model is tested critically against another over as wide a range of conditions and with as many different types of reaction as possible. It is probably true to say however that the 'perfect' model will not emerge until much more is known about the underlying physics of fluidization and in particular about the factors that govern the division of flow and the rate of exchange between the bubbles and the rest of the bed.

2.5 Multiple region models

There is widespread agreement with the conclusion of Chavarie and Grace, mentioned in the preceeding Section, that fluidized-bed reactor models take insufficient account of reaction in the entry and exit regions of the bubbling bed, that is to say, in the region close to the distributor and in the freeboard space above the bed surface. A number of authors have commented on the abnormally high rate of reaction near the distributor or grid where, in the embryonic stage of bubble development, interphase mass transfer appears to be very fast and gas–solid contact most pronounced. Similarly many observations have been made of freeboard-space temperature increases in the case of exothermic gas–solid reactions indicating an extra degree of reactant conversion in the region.

It would thus seem to be more realistic to model fluidized-bed reactors on the basis of three consecutive regions: the grid region, the bubbling bed itself and the freeboard. A number of attempts have been made to explore the consequences of this approach and it will be convenient to examine the treatments in each of the extreme regions in turn.

2.5.1 Reactions in the grid region

A relatively simple treatment of chemical reaction near the grid was proposed by Behie and Kehoe[77], who based their model on mass transfer between the fluidized emulsion phase and gas jets which were assumed to be formed at the point of entry into the bed of the fluidizing gas. The jets were assumed to be highly turbulent and to break up after a certain distance into conventional bubbles. Chemical reaction in the vicinity of the jets was described by means of the Davidson–Harrison model with a completely mixed dense phase. Mass transfer coefficients were measured for the jet region by Behie[78] using a 0.61 m diameter bed of 60 μm cracking catalyst and the values were used in the model predictions. Typical results are shown in *Figure 2.17* for both fast and slow reactions and it is clear that the grid model predicts a strong effect in cases where the kinetic rate constant is sufficiently fast for the overall conversion to be controlled by the hydrodynamics. Grace and de Lasa[79] have extended this model to more complex mixing situations in the grid region but as with the Behie–Kehoe approach the fundamental assumption made is that the fluidizing gas enters the bed in the form of jets.

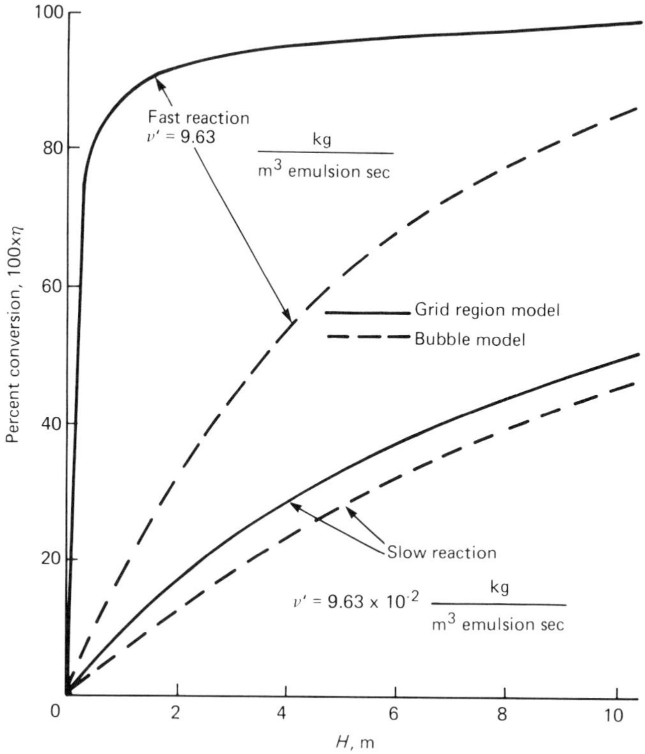

Figure 2.17 Effect of enhanced contacting efficiency near the distributor, for fast and slow reactions (From Behie and Kehoe[77])

The existence of such jets has been seriously brought into doubt however by the work of Rowe *et al.*[80] who used X-ray cine photography to examine the discharge of gas from an orifice into a three-dimensional gas-fluidized bed. They found in all cases and at velocities of up to 70 m/s that gas entered the bed as a stream of bubbles and that in no instance was a stable jet formed at the grid. It was concluded that the only circumstances under which a permanent jet-like structure was capable of being formed was if particles were locally defluidized or if a surface was available near the orifice on which a jet could stabilize. It was for this reason that jets were formed in two-dimensional[81] and semi-cylindrical[82] beds.

If indeed gas jets do not form at the grid of a three-dimensional fluidized bed then the observed enhancement of reaction rates in that region must be attributed to another cause and the mechanism of bubble formation at an orifice suggested in Chapter 1, Section 1.14 could provide an explanation. This model suggests that a region of expanded emulsion phase containing approximately half the bubble gas is produced around a bubble in the course of its formation and that the emulsion phase voidage only returns to the 'normal' value after the bubble has risen into the bed a distance equal to twice its diameter. During this period there is efficient contact between solid particles and at least half of the gas that finally appears in the bubble phase and, since the concentration of reactant at the point of entry to the bed is a maximum, then reaction can be expected to be efficient. The implications of the model have not yet been explored in depth but its mechanism could form the basis of a rational explanation of the enhanced reactivity of the grid region of a fluidized-bed reactor.

2.5.2 Reactions in the freeboard

When bubbles burst and release their gases at the surface of a fluidized bed it is generally observed that particles are ejected into the freeboard space above the surface. These particles can originate either from the dense phase just ahead of the bubble at the moment of eruption or from those solids that travel in the wake of the rising bubbles. Do *et al.*[83] from photographic observations of the surface of a two-dimensional bed reported that the ejected particles came from the nose of the bursting bubble, but a number of other workers, notably Kehoe[84], Basov *et al.*[85], Leva and Wen[86] and George and Grace[87] have given support to the alternative theory, and currently the weight of opinion seems to be heavily in favour of the bubble wakes as being the source of freeboard particles.

Now, in many cases a high proportion of the ejected particles quickly fall back to the surface and mix in with the rest of the bed material, the maximum height to which they are thrown being the 'transport disengaging height' or TDH (Chapter 1, Section 1.9). If however the fluidizing gas velocity is greater than their terminal fall velocity, and this is particularly likely in the case of the very finest material, the particles will be swept out of the bed and into the cyclones or whatever device is used to remove them from the exit gases. In either case there will be a concentration gradient of particles in the freeboard region and unreacted gases leaving the bed surface will have further opportunity for reaction with them.

In order to explore their possible significance a number of authors have proposed models of these freeboard reactions and in several cases experimental verification has been reported. Yates and Rowe[88] developed a very simple model of a catalytic reaction based on the assumption that the freeboard contained perfectly mixed, equally dispersed particles derived from bubble wakes. The fraction, f, of wake particles ejected was a model parameter and the two cases of plug flow and complete mixing of freeboard gas were considered.

The basic assumptions of the model are illustrated in *Figure 2.18* from which it may be seen that, on the basis of the two-phase theory, the rate of particle ejection from the surface is:

$$\frac{(1-\varepsilon_{mf})fQ_b}{3} = \frac{f(U-U_{mf})A_b(1-\varepsilon_{mf})}{3} \tag{2.106}$$

where Q_b is the volumetric flowrate of bubble gas and A_b is the bed

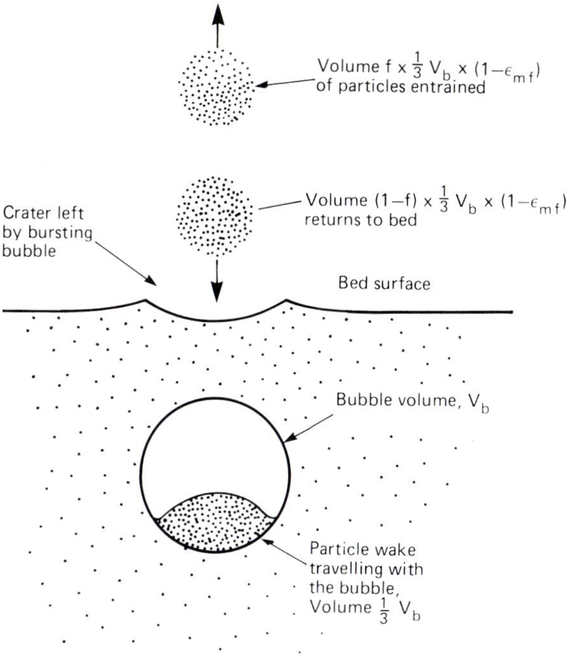

(a) *Ejection of wake particles into the freeboard*

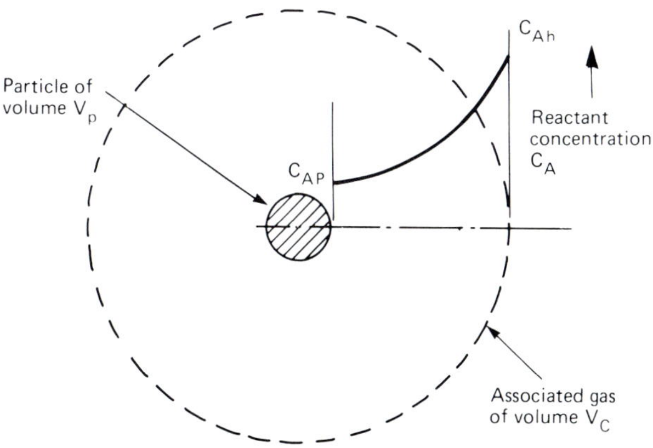

(b) *Concentration of reactant at the particle surface and at the outer surface of a gas cell.*

Figure 2.18 Physical basis of freeboard reaction model (From Yates and Rowe[88])

cross-sectional area. The particle hold-up in the freeboard, V, may then be expressed in terms of the freeboard height, H_F, and the terminal fall velocity, U_t, of the ejected particles:

$$V = \left(\frac{H_F}{U - U_t}\right) \frac{f(U - U_{mf})A_b(1 - \varepsilon_{mf})}{3} \tag{2.107}$$

giving the freeboard voidage ε_f as:

$$\varepsilon_f = 1 - \frac{f(1 - \varepsilon_{mf})}{3} \left(\frac{U - U_{mf}}{U - U_T}\right) \tag{2.108}$$

Now if the particles are assumed to be equally spaced (an approximation that may not be too extreme above the transport disengaging height) the volume, V_c, of the cell of gas surrounding each particle may be approximated by:

$$V_c = \frac{3V_p}{f(1 - \varepsilon_{mf})} \left(\frac{U - U_T}{U - U_{mf}}\right) \tag{2.109}$$

The rate of change of concentration of reactant gas A within each of these cells will be determined by the surface area of the particle, A_p, the concentration difference between the particle surface and the periphery of the cell, $C_{Ah} - C_{Ap}$, and a mass transfer coefficient, k_g:

$$-\frac{dC_A}{dt} = \frac{k_g A_p}{V_c} (C_{Ah} - C_{Ap}) \tag{2.110}$$

while the rate of change of concentration of A at the particle surface will be controlled by the rate of chemical reaction. For plug flow of reactant gas through the freeboard space the equations lead to the following expression for the fraction of gas unconverted at any height h above the surface of the bubbling bed:

$$\frac{C_{Ah}}{C_{As}} = \exp\left[-\frac{fh}{3(U - U_T)\left(\dfrac{d_p}{6k_g(1 - \varepsilon_{mf})} + \dfrac{1}{k}\right)}\right] \tag{2.111}$$

Here C_{As} is the concentration of reactant at the surface and k is an assumed first-order rate constant. The corresponding equation for the completely mixed case is:

$$\frac{C_A}{C_{As}} = \frac{1}{\left[1 + \dfrac{fH_F}{3(U - U_T)\left(\dfrac{d_p}{6k_g(1 - \varepsilon_{mf})} + \dfrac{1}{k}\right)}\right]} \tag{2.112}$$

A value for the mass transfer coefficient can be found from the expression of Rowe *et al.* (89):

$$Sh = 2 + 0.69Sc^{1/3}Re_t^{1/2} \tag{2.113}$$

$$\text{where the Sherwood number, } Sh = \frac{k_g d_p}{D_G} \tag{2.114}$$

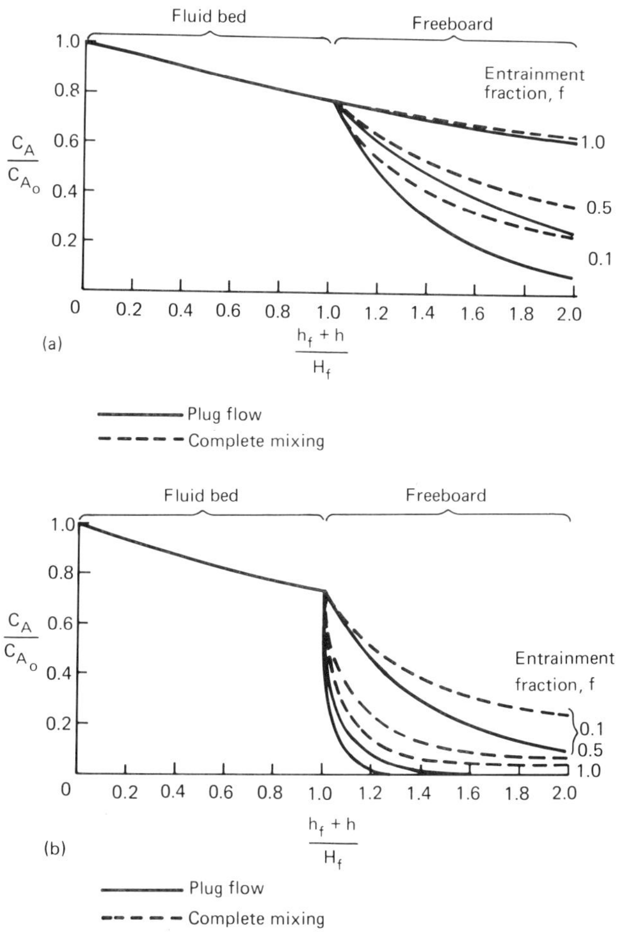

Figure 2.19 Predictions of freeboard reaction models (From Yates and Rowe[88])

$$\text{the Schmidt number, } Sc = \frac{\mu}{\rho D_G} \tag{2.115}$$

$$\text{and the terminal Reynolds number, } Re_t = \frac{U_t d_p \rho}{\mu} \tag{2.116}$$

Thus knowing the terminal fall velocity, U_t, Re_t may be obtained and from this the Sherwood number calculated via equation (2.113), giving k_g from equation (2.114). U_t may be obtained from published graphical correlations between Re_t and the velocity independent group $2d_p^3\rho g(\rho_s - \rho_g)/3\mu^2$ (Ref 90).

The model equations were solved for various combinations of the parameters (*Figure 2.19*) and it was shown that in many cases freeboard reactions lead to a greater conversion than that achieved by the fluid bed itself. The effect is expected to be most pronounced in systems operated at high gas velocity with a high proportion of catalyst fines. Furthermore since the effective thermal conductivity of the freeboard region is much less than that of the fluid bed itself, a rise in temperature is to be expected for exothermic reactions. Support for these contentions comes from the experimental studies of Miyauchi and his co-workers[91,92].

Further aspects of freeboard reactions will be discussed in Section 3.7.2.

References

1 GRACE, J.R., *A.I.Ch.E. Symp. Ser.*, **70**(141), 22 (1974)
2 DAVIDSON, J.F. and HARRISON, D., *Fluidised Particles*, Cambridge University Press (1963)
3 GRACE, J.R., *A.I.Ch.E. Symp. Ser.*, **67**(116), 159 (1971)
4 ROWE, P.N., *Proc. 2nd Int. Conf. Reaction Engng*, Amsterdam, A9 (1972)
5 PYLE, D.L., *Proc. 1st Int. Conf. Reaction Engng*, Washington, D.C., 106 (1970)
6 YATES, J.G., *Chem. Engineer*, 671 (1975)
7 SHEN, C.Y. and JOHNSTONE, H.F., *A.I.Ch.E. Jl*, **1**, 349 (1955)
8 MATHIS, J.F. and WATSON, C.C., *A.I.Ch.E. Jl*, **2**, 518 (1956)
9 LEWIS, W.K., GILLILAND, E.R. and GLASS, W., *A.I.Ch.E. Jl*, **5**, 419 (1959)
10 MAY, W.G., *Chem. Engng Prog.*, **55**(12), 49 (1959)
11 LANNEAU, K.P., *Trans. I. Chem. Engrs*, **38**, 125 (1960)
12 GOMEZPLATA, A. and SHUSTER, W.W., *A.I.Ch.E. Jl*, **6**, 454 (1960)
13 VAN DEEMTER, J.J., *Chem. Engng Sci.*, **13**, 143 (1961)
14 MAMURO, T. and MUCHI, I., *Int. Chem. Eng.*, **5**, 732 (1965)
15 KOBAYASHI, H., ARAI, F. and CHIBA, T., *Kagaku Kogaku*, **33**, 274 (1969)
16 MIREUR, J.P. and BISCHOFF, K.B., *A.I.Ch.E. Jl*, **13**, 839 (1967)
17 VAN DEEMTER, J.J., *Proc. Int. Symp. Fluidization*, Eindhoven; Netherlands University Press, 334 (1967)
18 ORCUTT, J.C., DAVIDSON, J.F. and PIGFORD, R.L., *Chem. Engng Prog. Symp. Ser.*, **58**(38), 1 (1962)
19 ROWE, P.N., in *Fluidisation*, Soc. Chem. Ind. London, 15 (1963)
20 ROSE, P.L., PhD Thesis, Cambridge University (1965)
21 PARTRIDGE, B.A. and ROWE, P.N., *Trans. I. Chem. Engrs*, **44**, 335 (1966)
22 TOOR, F.D. and CALDERBANK, P.H., *Proc. Int. Symp. Fluidization*, Eindhoven; Netherlands University Press, 373 (1967)
23 HOVMAND, S. and DAVIDSON, J.F., Trans. I. Chem. E, **46**, 190 (1968)

24 KUNII, D. and LEVENSPIEL, O., *Fluidization Engineering*, Wiley, New York (1969)
25 KATO, K. and WEN, C.Y., *Chem. Engng. Sci.* **24**, 1351 (1969)
26 CHIBA, T. and KOBAYASHI, H., *Chem. Engng Sci*, **25**, 1375 (1975)
27 FRYER, C.C. and POTTER, O.E., *A.I.Ch.E. Jl*, **22**, 38 (1976)
28 CHAVARIE, E. and GRACE, J.R., *Chem. Engng Sci.*, **31**, 741 (1976)
29 ZENZ, F.A. and OTHMER, D.F., *Fluidization and Fluid-Particle Systems*, Reinhold, New York (1960)
30 CALDERBANK, P.H. and TOOR, F.D., in *Fluidization*, Davidson, J.F.and Harrison, D., (eds), Chapter 8, Academic Press, London
31 HIGBIE, R., *Trans. Am. Inst. Chem. Eng.*, **31**, 365 (1935)
32 MURRAY, J.D., *J. Fluid Mech.*, **21**, 465 (1965)
33 MURRAY, J.D., *J. Fluid Mech.*, **22**, 57 (1965)
34 SIT, S.P. and GRACE, J.R., *Chem. Engng Sci.*, **33**, 1115 (1978)
35 SIT, S.P. and GRACE, J.R., *Chem. Engng Sci.*, **36**, 327 (1981)
36 TOEI, R., MATSUNO, R., NISHITANI, K., HAYASHI, H. AND IMAMATO, T., *Kagaku Kogaku*, **33**, 668 (1969)
37 PEREIRA, J.A.F., PhD Dissertation, Edinburgh University (1977)
38 YOSHIDA, K. and KUNII, D., *J. Chem. Eng.Japan*, **1**, 11 (1968)
39 GUPALO, Y.P., RYAZANTSEV, Y.S. and SERGEEV, Y.A., in *Fluidization*, Davidson, J.F. and Keairns, D.L., (eds), Cambridge University Press, 162 (1978)
40 DRINKENBURG, A.A.H. and RIETEMA, K., *Chem. Engng Sci.*, **28**, 259 (1973)
41 YATES, J.G. and CONSTANS, J.A.P., *Chem. Engng Sci.*, **28**, 1341 (1973)
42 NGUYEN, H.V. and POTTER, O.E., in *Fluidization Technology*, Keairns, D.L., (ed), Hemisphere Publishing, Washington, D.C., 290 (1976)
43 CAZES, R.C., MPhil Thesis, London University (1975)
44 BOHLE, H. and VAN SWAAIJ, W.P.M., in Fluidization, Davidson, J.F. and Keairns, D.L., (eds), Cambridge University Press, 167 (1978)
45 STEPHENS, G.K., SINCLAIR, R.J. and POTTER, O.E., *Powder Tech.*, **1**, 61 (1967)
46 LATHAM, R., HAMILTON, C. and POTTER, O.E., *Brit. Chem. Engr*, **13**, 666 (1968)
47 POTTER, O.E., in *Fluidization*, Eds. Davidson, J.F. and Harrison, D., Chapter 7, Academic Press, London (1971)
48 CHIBA, T. and KOBAYASHI, H., *Proc. Int. Symp. Fluidization and its Applications*, Toulouse, France, 468 (1973)
49 ROWE, P.N., *A.C.S. Symp. Ser.*, No. 65, 436 (1978)
50 ROWE, P.N., SANTORO, L. and YATES, J.G., *Chem. Engng Sci.*, **33**, 133 (1978)
51 DARTON, R.C., *Trans. I. Chem. Engrs*, **57**, 134 (1979)
52 ROWE, P.N. and PARTRIDGE, B.A., *Trans. I. Chem. Engrs*, **43**, 157 (1965)
53 ROWE, P.N., PARTRIDGE, B.A. and YATES, J.G., *Proc. Int. Symp. Fluidization*, Eindhoven; Netherlands University Press, Amsterdam, 711 (1967)
54 ELLIS, J.E., PARTRIDGE, B.A. and LLOYD, D.I., *Proc. Tripartite Chem. Engng Conf.*, Montreal, 43 (1968)
55 POTTER, O.E., *Catal. Rev.-Sci. Engng*, **17**(2), 155 (1978)
56 WOOLARD, I.N.M. and POTTER, O.E., *A.I.Ch.E. Jl*, **14**, 338 (1968)
57 YATES, J.G. and GREGOIRE, Y-Y., in *Fluidization*, Eds. Grace, J.R. and Matsen, J.M., 581, Plenum Press, New York (1980)
58 KOBAYASHI, H., ARAI, F. and CHIBA, T., *Chem. Engng Tokyo*, **29**, 858 (1965)
59 COOKE, M.J., HARRIS, W., HIGHLEY, J. and WILLIAMS, D.F., *Proc. Tripartite Chem. Engng Conf.*, Montreal, 14 (1968)
60 KOBAYASHI, H., ARAI, F. and SUNAWAGA, T., *Chem. Engng Tokyo*, **31**, 239 (1967)
61 TANAKA, T., MS Thesis, Hokkaido University (1967)
62 WERTHER, J., *Chem. Engng Sci.*, **35**, 372 (1980)
63 ASTARITA, G., *Mass Transfer with Chemical Reaction*, Elsevier, Amsterdam (1967)
64 DANCKWERTS, P.V., *Gas-liquid Reactions*, McGraw-Hill, London (1970)
65 WERTHER, J., *Trans. I. Chem. Engrs*, **52**, 160 (1974)
66 WERTHER, J., German Chem. Eng., **1**, 243 (1978)
67 MASSIMILLA, L. and JOHNSTONE, H.F., *Chem. Engng Sci.*, **16**, 105 (1961)
68 HEIDEL, K., SCHUGERL, K., FETTING, F. and SCHIEMANN, G., *Chem. Engng Sci.*, **20**, 557 (1965)
69 VAN SWAAIJ, W.P.M. and ZUIDERWEG, F.J., *Proc. 2nd Int. Symp. Reaction Engng*, Elsevier, Amsterdam, B9-25 (1972)

70 DE VRIES, R.J., VAN SWAAIJ, W.P.M., MANTOVANI, C. and HEIJKOOP, A., *Proc. 2nd Int. Symp. Reaction Engng*, Elsevier, Amsterdam, B9-59 (1972)
71 VAN SWAAIJ, W.P.M. and ZUIDERWEG, F.J., *Proc. Int. Symp. Fluidization and its Applications*, Toulouse, France, 454 (1973)
72 KRISHNA, R., Paper presented to Working Party on Chemical Reaction Engineering, European Federation of Chemical Engineering, Ghent, Belgium, September (1979)
73 CHAVARIE, C. and GRACE, J.R., *Ind. Engng Chem. Funds.*, **14**, 75, 79, 86 (1975)
74 SHAW, I.D., HOFFMAN, T.W. and REILLY, P.M., *A.I.Ch.E. Symp. Ser.*, **70**,(141), 41 (1974)
75 HOVMAND, S. and DAVIDSON, J.F., in *Fluidization*, Eds. Davidson, J.F. and Harrison, D., Chapter 5, Academic Press, London (1971)
76 RAGHURAMAN, J. and POTTER, O.E., *A.I.Ch.E. Jl*, **24**, 698 (1978)
77 BEHIE, L.A. and KEHOE, P., *A.I.Ch.E. Jl*, **19**, 1070 (1973)
78 BEHIE, L.A., PhD dissertation, University of Western Ontario (1972)
79 GRACE, J.R. and DE LASA, H.I., *A.I.Ch.E. Jl*, **24**, 364 (1978)
80 ROWE, P.N., MACGILLIVRAY, H.J. and CHEESMAN, D.J., *Trans. I. Chem. Engrs*, **57**, 194 (1979)
81 ZENZ, F.A., *Proc. Tripartite Chem. Eng. Conf.*, Montreal, 136 (1968)
82 YANG, W.C. and KEAIRNS, D.L., in *Fluidization*, Eds. Davidson, J.F. and Keairns, D.L., 208, Cambrdige University Press (1978)
83 DO, H., GRACE, J.R. and CLIFT, R., *Powder Tech.*, **6**, 195 (1972)
84 KEHOE, P.W.K., PhD Thesis, Cambridge University (1969)
85 BASOV, V.A., MARKHEVKA, V.I., MELIK-AKHNAZAROV, T.K. and OROCHKO, D.I., *Int.Chem. Engng*, **9**, 263 (1969)
86 LEVA, M. and WEN, C.Y., in Fluidization, Eds Davidson, J.F. and Harrison, D., Chap 14, Academic Press, London (1971)
87 GEORGE, S.E. and GRACE, J.R., *A.I.Ch.E. Symp. Ser.*, **74**(176), 67 (1978)
88 YATES, J.G. and ROWE, P.N., *Trans. I. Chem. E.*, **55**, 137 (1977)
89 ROWE, P.N., CLAXTON, K.T. and LEWIS, J.B., *Trans. I. Chem. E.*, **43**, 14 (1965)
90 COULSON, J.M. and RICHARDSON, J.F., 'Chemical Engineering', Vol. 2, Pergamon Press, Oxford, page 490 (1955)
91 MIYAUCHI, T. and FURUSAKI, S., *A.I.Ch.E. Jl*, **20**, 1087 (1974)
92 FURUSAKI, S., KIKUCHI, T. and MIYAUCHI, T., *A.I.Ch.E Jl*, **22**, 354 (1976)

CHAPTER 3

Catalytic cracking

3.1 Process developments

The objective of cracking is to convert heavy petroleum distillates into lighter fractions boiling in the gasoline and middle oil ranges. The earliest process for this–thermal cracking–achieved the breakdown of the larger distillate molecules simply by the application of heat, but the general lack of adequate product quality control and the relatively poor engine performance of the gasolines it produced were serious disadvantages of the method. Work carried out by the French engineer Eugene Houdry in the 1920s showed that better quality gasoline could be made by treating heavy distillates at high temperatures with acidified montmorillonite clays and this led to the introduction in 1936 of industrial catalytic cracking. In the subsequent forty years the technique has been developed into what is certainly the most important fluidized bed process, and probably the most important catalytic process, in all industry.

One of the central problems in catalytic cracking is the rapid deactivation of the catalyst caused by the surface deposition of a solid hydrocarbon polymer of high molecular weight generally known as 'coke'. Coke originates mainly in two ways, (i) catalytically via the formation of condensed polynuclear hydrocarbon residues on the catalyst surface, (ii) thermally via homogeneous gas-phase pyrolyses that are characterized by the Conradson carbon number. A third type of carbon, 'additive carbon', may be formed for some charge stocks–its origin is unknown but its formation seems to be linked to the basic nitrogen content and average molecular weight of the oil[1]. Catalyst regeneration is achieved by burning off the coke deposits with air.

The first large-scale catalytic cracking process, the Houdry process, used several fixed bed reactors connected in parallel so that continuous operation could be maintained while individual beds were undergoing a sequence of reaction, purging, regeneration and further purging. The drawbacks of this intermittent mode of operation are obvious, and it was in an effort to overcome them and at the same time to improve the yield and quality of the gasoline product, that first the moving bed and then the fluidized bed catalytic cracking (FCC) processes were developed. The first

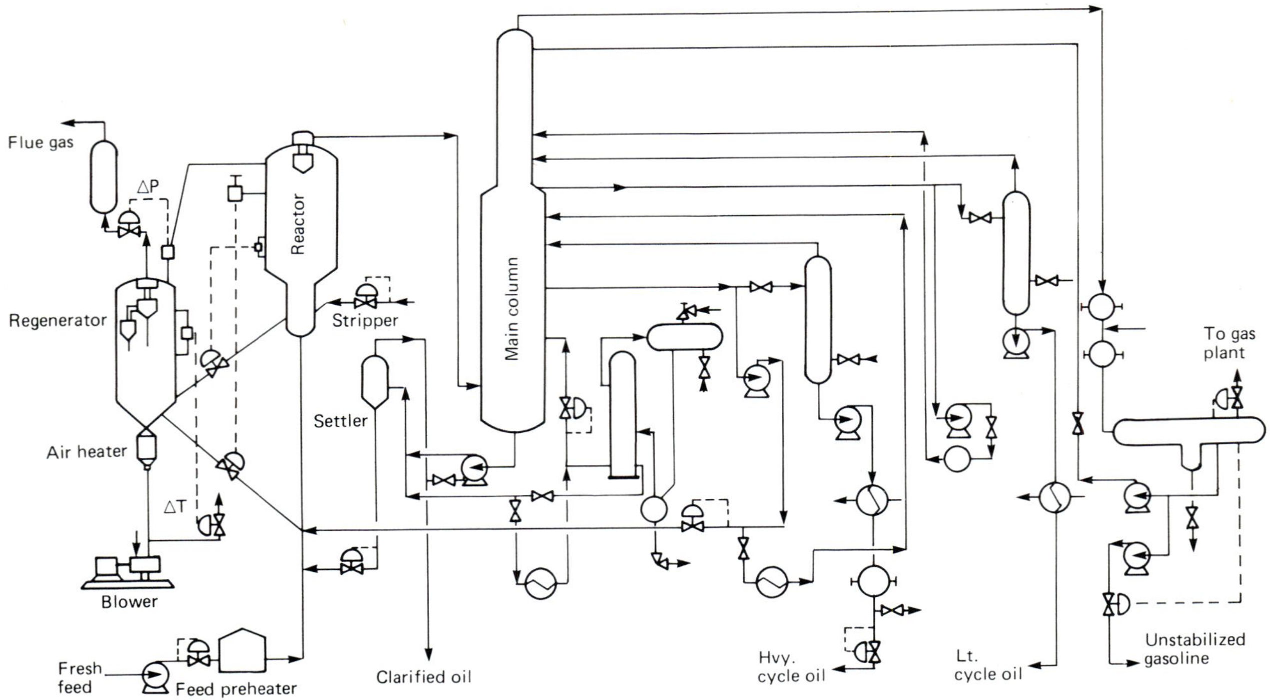

Figure 3.1 UOP-designed FCC unit, circa 1963 (From Pohlenz[4])

FCC plant started operation at the Baton Rouge refinery of the Standard Oil Co. in 1942[2].

The next twenty years saw many improvements in engineering design, most of which were aimed at reducing both the size of the reactor units and the amount of ancilliary equipment required[3]. Despite the considerable variation in reactor configuration that developed during this period the general principle in all the designs is the same (*Figure 3.1*). Cracking is carried out in one vessel where the hot catalyst is fluidized with vaporized feed. Deactivated catalyst flows from the reactor through either a standpipe or a riser to the regenerator where it is fluidized with air and the coke deposit is burned off. The reduction in catalyst temperature caused by the endothermic cracking reaction is counteracted by the exothermicity of the regeneration step and, by use of this so-called 'thermal flywheel', benefit is gained from the otherwise disadvantageous loss of catalyst activity caused by the coking reaction. Before passing to the regenerator the catalyst flows through a steam stripping section where any adsorbed volatiles are removed. Cracked products and steam leave from the top of the reactor and pass to a fractionator; steam and the oxides of carbon, nitrogen and sulphur leave from the regenerator. Catalyst attrition cannot be prevented entirely and both reactor and regenerator are fitted with cyclones. By controlling the flowrates of feed, air and steam a continuous flow of catalyst is maintained between reactor and regenerator.

The flow of hot catalyst into the reactor is controlled by a slide valve in the transfer line while the spent-catalyst slide valve is operated in response to a level controller in the reactor. Pressure in the reactor is controlled by means of the gas compressor in the recovery system, regenerator pressure being controlled by a pressure difference instrument located between reactor and regenerator.

An interesting variant on this basic theme, and one which was later to undergo considerable development, is the two-stage unit with a riser reactor designed by Shell[5]. This evolved from observations that conversion efficiences were higher when the feed vapour was introduced into a long riser than when it was injected directly into the bed; the final configuration was engineered so that the fresh feed was cracked in the riser (which thus acted as a transported-bed reactor) and recycle stock after product separation was fed to a conventional reactor.

Since the early 1960s great changes have been brought about by the introduction of the ultra-high activity zeolite catalysts. In addition to their greater activity the zeolites have a higher selectivity than the conventional catalysts for the production of gasoline, typically 84 vol. per cent compared with 71 vol. per cent[6]. It has been estimated that since their introduction on a large scale in 1962 this enhanced selectivity has resulted in the conservation of about 200 million barrels per year of crude oil in North America alone with resultant economic savings of $250 to $350 million per year[6]. These catalysts and the effects they have had on design and operation of FCC units are discussed below.

Fluidized catalytic cracking is a very flexible process and can accommodate a wide range of feedstocks from straight run distillates to heavy cycle oils. It is more common however for FCC feedstock to be taken from the vacuum distillation unit as a material containing low concentrations of catalyst fouling agents such as metals, (Ni and V), heavy asphaltines and heterocyclic compounds. Both virgin and coker gas oils are also common feeds although the former are the more reactive.

3.2 Riser cracking

A modern configuration utilizing the high cracking activity of zeolite catalysts is shown in *Figure 3.2*. Cracking occurs in the vertical riser where

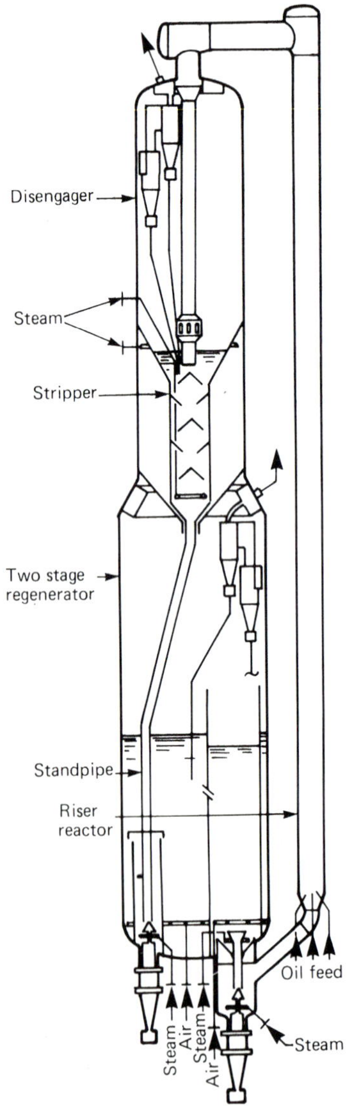

Figure 3.2 A modern riser cracker

oil vapour and regenerated catalyst flow cocurrent upwards. Risers can be up to 45 m in height. Separation of catalyst and products takes place in a disengaging zone at the top of the riser. Residual volatiles are removed from the catalyst in a steam stripping section from which the catalyst flows through a standpipe to the regenerator below. Catalyst circulation is controlled by maintaining density differences around the circuit through regulation of air, oil vapour and steam flows. Typical densities quoted in the early literature[7] are: regenerator bed 420–560 kg/m^3; catalyst standpipes 560–760 kg/m^3: riser and lift lines 34–84 kg/m^3.

The processes occuring in the riser are made complicated by the changing conditions along its length–the vapour volume increases as cracking proceeds but the rate of increase slows as the catalyst activity is reduced by the coking reaction. Ideally the catalyst should be in plug flow with a minimum of backmixing although gross departures from ideality can occur in practice. For example, a Shell publication[8] gives results of tests made on a modern commercial unit in which considerable variations in conversion were observed (*Figure 3.3*). The high yield of product near to

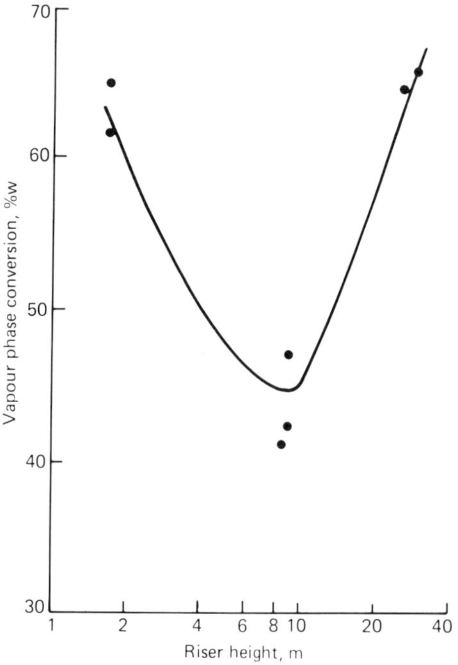

Figure 3.3 Typical vapour-phase conversion profile (From Schuurmans[8])

the feed injection point was attributed to a partial vapourization of the feed at that point on contact with hot catalyst and a subsequent rapid reaction. The part of the feed unvapourized and hence unreacted was assumed to have been poorly distributed and to have contacted the catalyst only much later in its passage through the reactor.

In normal operation the velocity of the solids is less than that of the vapours, the difference being the slip velocity which after the initial acceleration at the inlet point approaches the terminal fall velocity of the solids, U_t. The slip ratio, S, is defined as:

$$S = \frac{t_c}{t_o} = \frac{U_o}{U_o - U_t} \tag{3.1}$$

where U_o is the oil vapour velocity and t_c and t_o are the catalyst and oil residence times respectively. Clearly the slip ratio decreases with increasing U_o and approaches unity for $U_o \gg U_t$. In practice, oil velocities of up to 20 m/s are used with solids whose terminal fall velocities are 0.6–1.2 m/s[9].

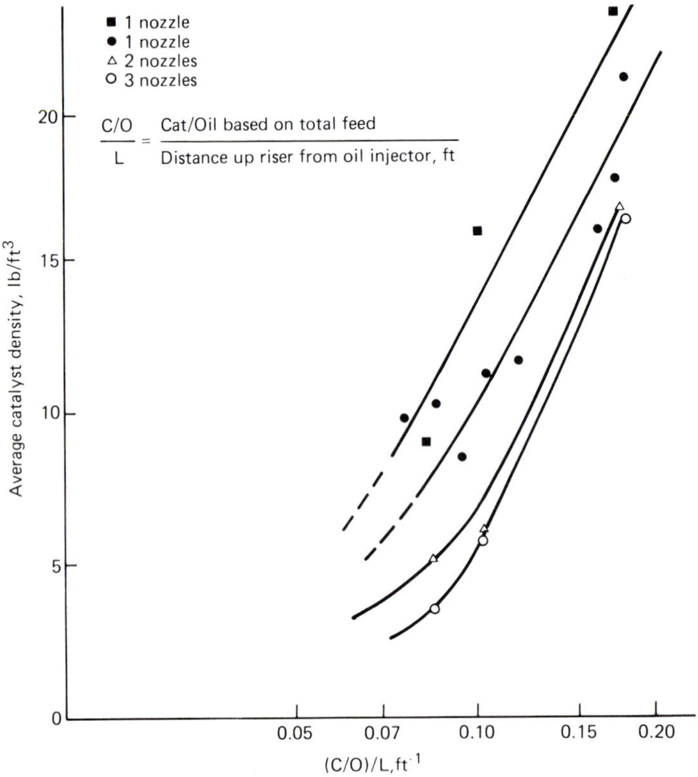

Figure 3.4 Axial distribution of catalyst density in a riser cracker (From Saxton and Worley[10])

High gas velocities also serve to improve the radial distribution of catalyst particles across the riser, although proper design of the oil injection system is probably of greater importance in this respect. Tests on a commercial riser have been reported in which catalyst distributions both axially and radially were measured using radioactive tracers[10]. Axial distributions are shown in *Figure 3.4* as a function of the number of oil feed nozzles used. Local densities of over 1000 kg/m^3 were measured near the walls of the reactor, but at the centre values as low as 50 kg/m^3 were found. Similar results have been reported from tests on a modern riser reactor operated with a feedrate of 3000 tons/day and in which catalyst density profiles were measured using a small radioactive source (^{137}Cs) that could be moved horizontally inside the riser[8]. Typical values obtained are shown in *Figure 3.5*, although no data are given on the actual dimensions of the unit.

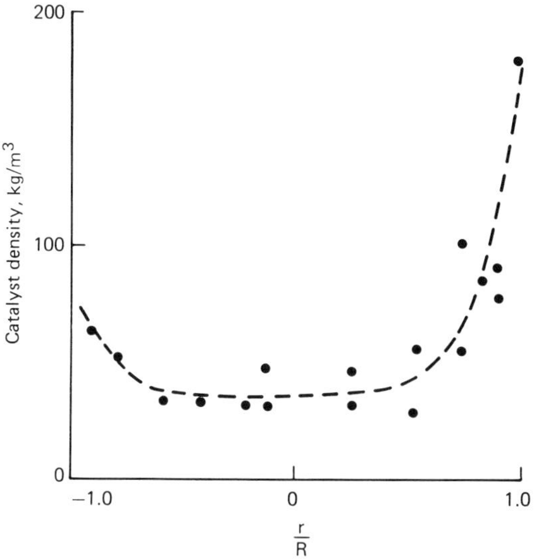

Figure 3.5 Typical radial catalyst density profile in feed riser (From Schuurmans[8])

In order to maximize catalyst distribution and to prevent the formation of zones of high density, multiple high-velocity injection points are used to introduce the oil vapour with further assistance being provided by dispersion steam[10].

Experience has shown that cracking at higher temperatures raises the octane number of the gasoline product[9] and modern systems are designed to operate at 540°C and higher.

Corresponding to these higher reactor temperatures there has been a general move in recent years towards higher regenerator temperatures and a design value of around 675°C is now used[9]. This is desirable in view of the strong temperature dependence of the coke-burning reaction, (Pansing[11] gives a value of 146 kJ/mol for the activation energy of coke combustion) and the need to reduce the zeolite coke content to a low value. However the active life of zeolites is shortened by exposure to high temperatures for long periods of time and so efforts are made to reduce as much as possible the hold-up of solids in the regenerator. One method in widespread use is to carry out the regeneration in two stages[9]. As shown in *Figure 3.6* a minimum catalyst inventory, some 60 per cent of that for single-stage operation in which carbon content is reduced from 0.85 to 0.05 wt per cent, results from a 4:1 split between the two stages[9].

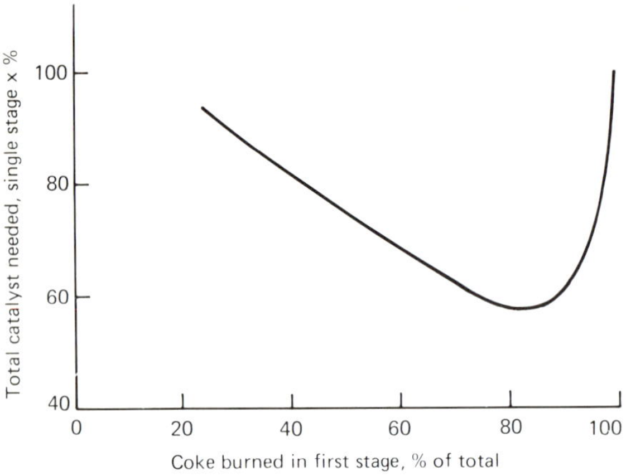

Figure 3.6 Two-stage regenerator inventory v. division of burning (From Whittington *et al.*[9])

Coke burning rates are also dependent on the dense phase partial pressure of oxygen, a quantity that is determined by the rate of interphase diffusion between bubbles and emulsion. This matter is discussed in detail below (Section 3.7.2), but it is worth noting here that in order to minimize afterburning of CO in the freeboard region of the regenerator, the airflow into the bed is maintained carefully at a value close to the stoichiometric requirement (11.5 kg of air/kg catalyst coke), oxygen concentrations in the flue gases being kept at well below one per cent[4]. Good distribution of fluidizing air is crucial in ensuring that these criteria are maintained.

3.3 Catalysts

The first cracking catalysts to be used industrially were naturally-occurring clays such as montmorillonite. These were gradually superseded by synthetic clays–the amorphous silica-aluminas–which in turn have been largely replaced by the synthetic crystalline zeolites. The latter account for over 90 per cent of all cracking catalysts currently in use.

The natural clays have two main disadvantages in use: they are deactivated rather too readily by sulphur compounds and they are sensitive to high regeneration temperatures. Although improvements were made in both these respects the materials were gradually replaced by the generally superior synthetic silica-aluminas. These are produced by mixing hydrogels of silica and alumina under carefully controlled conditions of pH, temperature and concentration. The resulting gel is filtered, washed, slurried with water and spray dried to give a finely divided, amorphous material with a surface area of the order of 500 m^2/g. Two broad classes have been employed commercially, the 'low alumina' type containing 10–15 per cent Al_2O_3 and the 'high alumina' with 20–30 per cent Al_2O_3. The chemistry of both natural clay and synthetic silica-alumina catalysts has been comprehensively reviewed by Ryland *et al.*[12].

The zeolites are porous, crystalline aluminosilicates many of which have the general formula $M_{x/n}[(AlO_2)_x(SiO_2)_y].zH_2O$ where x/n is the number of exchangeable cations, M, of valency *n*. About forty different types of zeolite are known to occur naturally[13] but only one, the rare mineral faujasite, is relevant to catalytic cracking. The basic unit of the faujasite structure is the cubo-octahedron made up of twenty four tetrahedra of either SiO_4^{4-} or AlO_4^{5-}. These are connected through their hexagonal faces to give an assembly of 192 tetrahedra in a large cubic unit cell with an edge

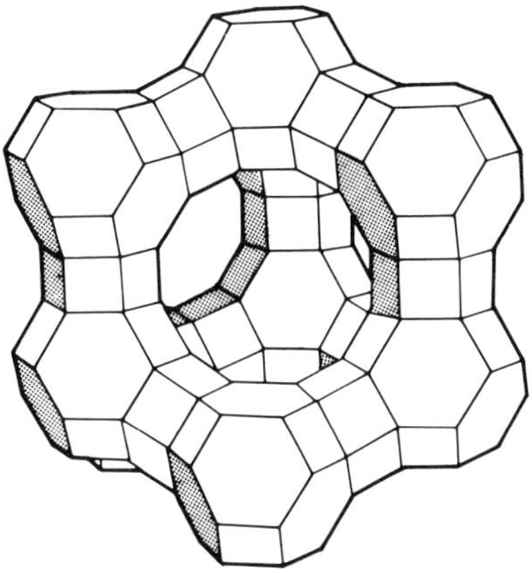

Figure 3.7 Arrangement of cubo-octahedra in the framework structure of faujasite

length of 24.95°A (*Figure 3.7*). Its open, cage-like structure with large cavities interconnected by channels makes the crystal porous and enables it to be penetrated by molecules with cross-sections smaller than that of the channels. The name 'molecular sieve' has been applied to those zeolites with very narrow channels.

The synthetic X and Y zeolites have the faujasite structure but are chemically different from it[14]. The general formulae of some synthetic zeolites are given in *Table 3.1*.

Table 3.1 (From Breck *et al.*[15])

Zeolite	*Chemical analysis*	*Unit cell composition*	*Ref.*
A	$Na_2O, Al_2O_3, 2SiO_2, 4.5H_2O$	$Na_{96}[(AlO_2)_{96}(SiO_2)_{96}]216H_2O$	16, 17
X	$Na_2O, Al_2O_3, 2.5SiO_2, 6H_2O$	$Na_{86}[(AlO_2)_{86}(SiO_2)_{106}]264H_2O$	18
Y	$Na_2O, Al_2O_3, 5SiO_2, 8H_2O$	$Na_{56}[(AlO_2)_{56}(SiO_2)_{136}]250H_2O$	19

Normally they are made in the form of the sodium compound, other cations (NH_4^+, Ca^{2+}, La^{3+}, etc.) being substituted for Na^+ as required by ion-exchange. Zeolite X is made by crystallizing a sodium alumino silicate gel prepared by mixing aqueous solutions of sodium aluminate, sodium silicate and sodium hydroxide, the latter in excess of that required by the stoichiometry. The amorphous gel so formed is agitated and heated at 100°C until crystallization is complete[18]. Zeolite Y is made in a similar way, the main differences in method being the addition of colloidal silica to the mixture and the 24-hour room temperature ageing of the gel carried out before crystallization[19]. Zeolite Y is much more stable to heat and acid than zeolite X and has very different catalyst support properties.

Of great importance from a catalytic point of view is the effective pore diameter of the crystals. For the sodium X zeolite it is about 9 to 10Å and is reduced by calcium exchange to 8 to 9 Å; the sodium Y zeolite and its calcium exchanged form have effective pore diameters of 9 to 10 Å .

The catalytic cracking activity of some ion-exchanged zeolites is several orders of magnitude greater than that of the amorphous silica-aluminas. *Table 3.2* shows results of experiments in which *n*-hexane was cracked over

Table 3.2 Catalytic activity of various zeolites compared with silica-alumina in *n*-hexane cracking (from Miale *et al.*[20])

Catalyst	*Major cations exchanged into zeolite*	*Analysis of zeolite, wt %*					*Activity*[a] *°C*	α[b]
		SiO_2	*Al_2O_3*	*Na*	*Ca*	*RE*		
Amorphous SiO_2–Al_2O_3 (standard catalyst)		...	...	...	...	...	540	1.0
	Ca^{2+}	47.8	31.5	7.7	12.3	...	530	1.1
Faujasite	NH_4^+	75.7	23.1	0.4	...	...	350	6400
	La^{3+}	...	...	0.4	...	29.0	270	7000
	RE	...	...	0.39	...	28.8	<270	>10000
	RE, NH_4^+	40.0	33.0	0.22	...	26.5	<270	>10000
	RE, NH_4^+	...	...	...	...	...	420	20
Zeolite A	Ca^{2+}	42.5	37.4	7.85	13.0	...	560	0.6
Zeolite ZK5	...	...	...	...	...	...	400	38
	H^+	76.8	23.1	0.47	...	...	340	450
Mordenite	Ca^{2+}	(~77)	...	1.01	...	...	520	1.8
	Ca^{2+}, H^+	82.0	14.0	0.4	...	...	360–400	40–200
	NH_4^+	...	...	...	...	...	<270	>10000
	H^+	80.1	13.4	0.3	1.54	...	300	2500
	NH_4^+	...	...	0.1	...	...	<270	>10000
Gmelinite	NH_4^+	...	...	...	...	...	270(~200)	>10000
Chabazite	NH_4^+	...	...	...	...	...	<270	>10000
Stilbite	NH_4^+	...	...	...	...	...	370	120
Offretite	NH_4^+	...	...	...	...	...	<270	>10000

[a]Temperature required to achieve 5 to 20% conversion of *n*-hexane.
[b]Activity of zeolite extrapolated to 540°C divided by the activity of silica-alumina at 540°C.

a number of catalysts and their activities per unit catalyst volume were compared, the activity of a standard silica-alumina being set at unity[20]. It has not been possible however to utilize this enhanced activity fully on the commercial scale, since some of the more desirable features of the FCC process (for instance the heat transfer function of the fluidized catalyst) make it necessary to have a relatively large solids hold up in the reactor–regenerator system. Their activity has been reduced by incorporating small particles of zeolite into a less active base such as amorphous silica-alumina. A typical cracking catalyst contains from 3 to 25 weight per cent of 1 μm diameter crystallites of zeolite embedded in a silica-alumina matrix[21]. High zeolite loadings give high gasoline yields (summer operation) while low zeolite loadings give high fuel oil yields (winter operation)[22].

Zeolites containing trivalent rare earth cations such as La^{3+}, Ce^{3+}, Pr^{3+}, Nd^{3+}, Sm^{3+} are extremely stable at high temperatures and in the presence of steam, a property which makes them very suitable for the FCC process [22]. Thus rare earth exchanged zeolites containing less than 0.5 weight per cent Na_2O will withstand high steam concentrations up to 760°C and thermal treatment up to 815°C[23]. Of even greater importance than the increased activity of the zeolites compared with silica alumina is their greater selectivity to gasoline and their correspondingly low yields of C_1–C_4 gases and coke (*Table 3.3*). This is attributed to their higher activity

Table 3.3 Comparison of yield structure for fluid catalytic cracking of waxy gas oil over commercial equilibrium zeolite and amorphous catalysts[a] (From Venuto and Habib[1])

Yields, at 80 vol% conversion[b]	*Amorphous, high alumina*	*Zeolite, XZ-25*	*Δ change from amorphous*
Hydrogen, wt%	0.08	0.04	–0.04
$C_1 + C_2$s, wt%	3.8	2.1	–1.7
Propylene, vol%	16.1	11.8	–4.3
Propane, vol%	1.5	1.3	–0.2
Total C_3s	17.6	13.1	–4.5
Butenes, vol%	12.2	7.8	–4.4
i-Butane, vol%	7.9	7.2	–0.7
n-Butane, vol%	0.7	0.4	–0.3
Total C_4s	20.8	15.4	–5.4
C_5-390 at 90% ASTM gasoline, vol%	55.5	62.0	+6.5
Light fuel oil, vol%	4.2	6.1	+1.9
Heavy fuel oil, vol%[c]	15.8	13.9	–1.9
Coke, wt%	5.6	4.1	–1.5
Gasoline octane no., R+O	94	89.8	–4.2

[a]Data from Blazek[24], run at 510°C and 1 bar in a fixed fluid bed.
[b]No loss basis on feed.
[c]HFO = 100 – conversion – LFO.

for hydrogen transfer relative to cracking due, it is thought, to the enhanced potential for bimolecular reactions of the hydrogen transfer type in the small pore structure of the zeolite crystallites[21]. The weaker acid sites on the zeolite surfaces are also of importance in this context.

3.4 Process chemistry

3.4.1 Cracking

The cracking ot petroleum hydrocarbons over acidic zeolite or silic-alumina catalysts proceeds in the main by endothermic reactions involving carbonium-ion intermediates, and the products formed depend on the structures of the individual components of the oil. Thus alkanes crack to produce alkenes and smaller alkanes, alkenes to produce smaller alkenes and cycloalkanes (naphthenes). Aromatic compounds with alkyl side chains are either dealkylated completely to an unsubstituted aromatic and an alkene, or partially cracked to an alkane and an alkenyl aromatic. The aromatic nuclei themselves are difficult to crack under normal conditions. Once these primary processes have taken place other secondary reactions occur and lead to the final products. Most important among these are hydrogen-transfer reactions, say from a naphthene to an alkene giving an aromatic and an alkane, isomerizations in which iso-alkanes are formed from their linear counterparts, and condensation reactions of aromatic residues with alkenes leading to complex polynuclear hydrocarbons that adhere to the catalyst surface. These latter compounds then undergo aggregation to form coke[25].

Not only the nature of the resulting products but also the rate and extent of cracking reactions are determined by the molecular structure of the reactants. Thus, in general, bonds to tertiary carbon atoms crack more readily than those to secondary carbon atoms while species with primary carbons are even less reactive[26]. This is a reflection of the greater ease of formation of carbonium ions at tertiary carbon centres and the corresponding facility of β–scissions once the ionic intermediate has been formed. Of the non-aromatic species, normal alkanes are less readily cracked than naphthenes with the same carbon number; the corresponding mono-alkyl aromatics are more reactive than either and crack readily to a high level of conversion[27].

The cracking of a linear alkane may be summarized as follows. Reaction is initiated by an interaction between an adsorbed hydrocarbon molecule and a proton derived from the surface of the solid catalyst:

(i) 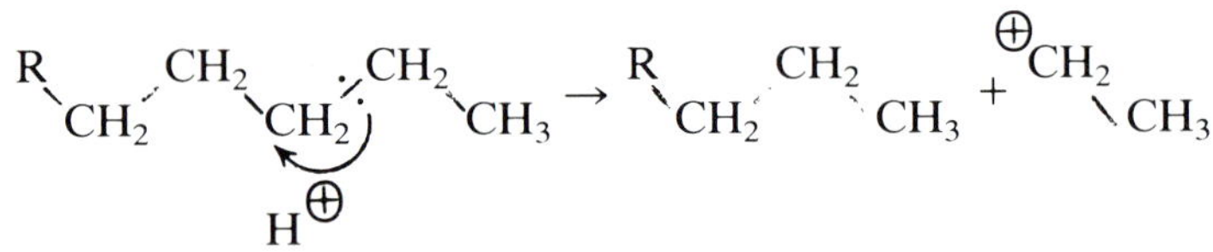

Chain propagation follows by the ethyl carbonium ion abstracting a hydride ion from a second alkane molecule:

(ii) 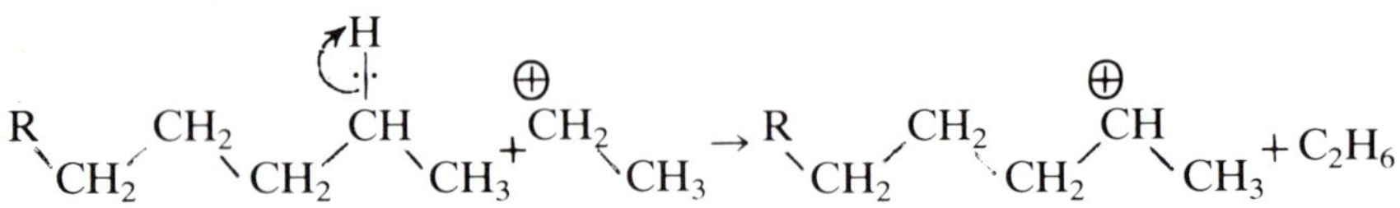

The secondary carbonium ion so formed undergoes a β–scission reaction to form a primary carbonium ion and an alkene:

(iii) 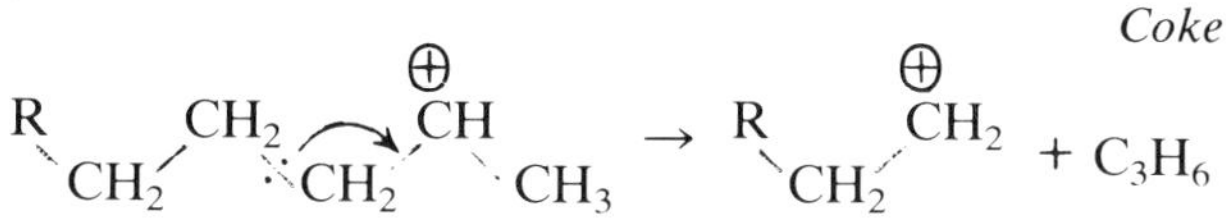

The primary carbonium ion formed in (iii) may propagate the chain by abstracting a hydride ion from another alkane molecule as in (ii) or it may rearrange to form a more stable secondary ion:

(iv) 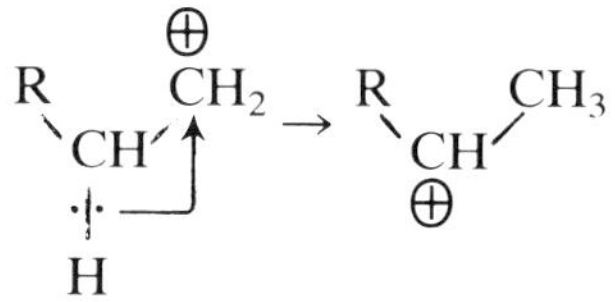

which may then react as before. Chain termination occurs through donation of a proton back to the catalyst surface:

(v) 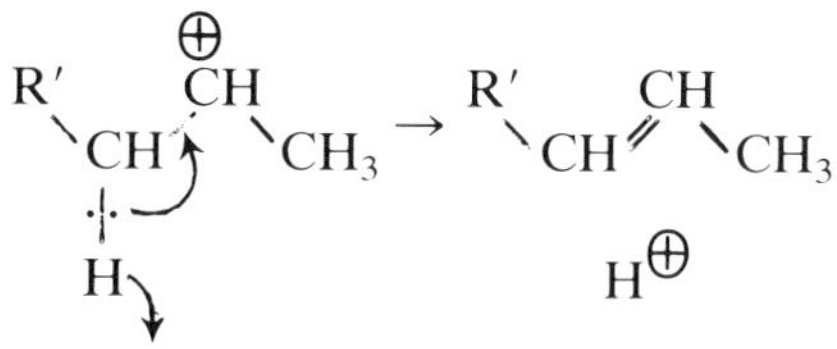

Reactions of cycloalkanes follow a similar pattern.

The chemistry of catalytic cracking reactions was reviewed extensively some years ago by Voge[27] and by Haensel[28]; more recent summaries are to be found in Gates *et al.*[21] and in Venuto and Habib[1].

3.4.2 Coke formation

The details of the chemical pathways leading to the formation of coke are less well understood than are those of the cracking reaction itself, and this is due partly to the ill-defined nature of the product. Coke on a spent catalyst is a complex mixture of semi-liquid polynuclear aromatic hydrocarbons derived from naphthalene, phenanthrene, chrysene etc., and a graphite-like solid. It would appear therefore that coke formation proceeds through a series of condensation–dehydrogenation reactions involving aromatic or alkenic nuclei which, given sufficient time, leads to the formation of a solid precursor of graphite. Appleby *et al.*[25] carried out a comprehensive study of the reactions of a number of hydrocarbons and heterocyclic compounds over a commercial silica-zirconia-alumina cracking catalyst and found the aromatics and alkenes to have the strongest tendencies to form coke, with carbonium ions being the most likely intermediates. They found a strong correlation between the basicity[29] and the coke-forming tendency of a number of aromatics and attributed this to the increasing probability of carbonium ion formation from adsorbed species (*Figure 3.8*). Furthermore the tendency to form coke was found to

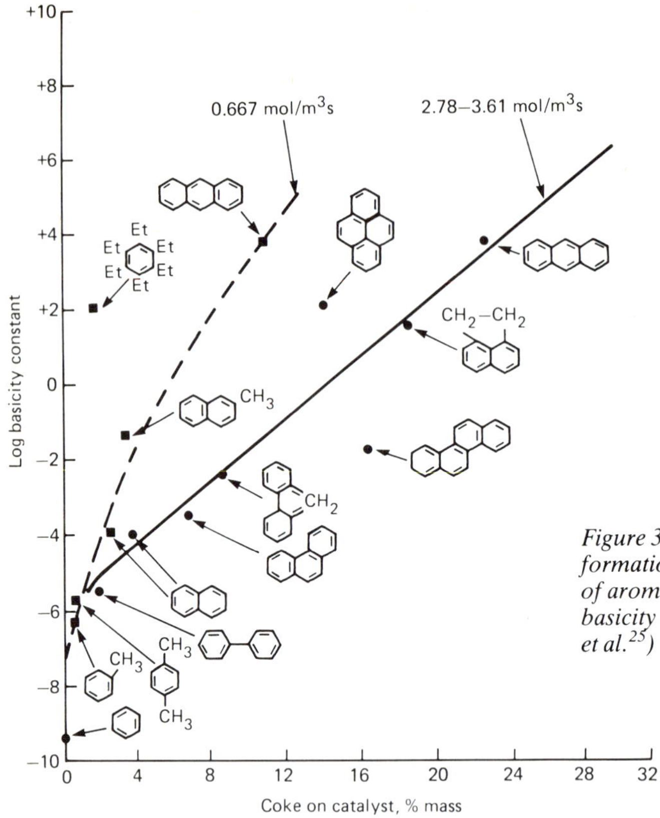

Figure 3.8 Coke formation as a function of aromatic hydrocarbon basicity (From Appleby et al.[25])

increase with increasing substitution of the aromatic nuclei in all cases except hexaethylbenzene where coke formation was quite low. The formation of coke from alkenes has been studied extensively[1,21,30–35] and it would appear that these compounds, when exposed to acidic surfaces, cyclize via hydrogen-transfer processes to generate the aromatic nuclei which then condense in the manner suggested above. Shephard *et al.*[31] concluded that the cyclization step was rate determining in the polymerization of propylene over silica–alumina at 140–392°C.

3.5 Energy considerations

The independent variables that define the reactor–regenerator operating conditions are reactor temperature, feed temperature, space velocity, catalyst activity and reactor pressure. In older designs of FCC unit where recycle was employed the combined ratio of fresh feed to recycled feed was an additional independent variable but in modern practice recycle is largely omitted[9]. Another change from traditional practice is the discontinuation of water spraying in the freeboard of the regenerator to control the temperature excursions caused by afterburning of carbon monoxide. The high temperature alloy steels in current use permit of higher regenerator off-gas temperatures[10], but careful control of the excess air for coke combustion (0–0.2 per cent) and the use of controlled-combustion catalysts with a high activity for CO oxidation in the regenerator bed, have reduced the afterburning problem to small proportions[1].

The most important dependent variables of the system are regenerator temperature, catalyst-to-oil ratio (i.e. catalyst circulation rate) and overall conversion. In the normally adiabatic mode of operation the combustion of coke laid down on the catalyst in the cracking stage provides sufficient heat for the total requirements of the system. Thus a simplified heat balance on the reactor, ignoring both the effects of steam injection in the stripping section and losses due to conduction and radiation, gives, following Kunii and Levenspiel[36]:

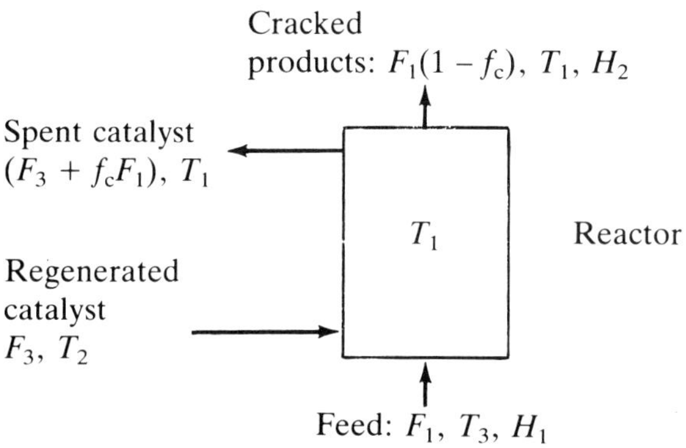

where F_1 = flow rate of feed (kg/s)
F_3 = flowrate of solid catalyst (kg/s)
f_c = mass fraction of feed deposited as coke
T_1 = reactor temperature (°C)
T_2 = regenerator temperature (°C)
T_3 = feed temperature (°C)
H_1 = feed enthalpy (kJ/kg)
H_2 = product enthalpy (kJ/kg)

Hence: (heat lost by catalyst) = (heat of cracking reaction) + (heat gained by feed)

$$F_3C_{p,s}(T_2 - T_1) = F_1\Delta H_{crack.} + [F_1(1-f_c)]H_2 - F_1H_1 \qquad (3.2)$$

$$\text{i.e.} \quad \frac{F_3}{F_1} = \frac{\Delta H_{crack.} + (1-f_c)H_2 - H_1}{C_{p,s}(T_2 - T_1)} \qquad (3.3)$$

where $C_{p,s}$ = specific heat capacity of solid catalyst (kJ/kg°C)
ΔH_{crack} = heat of endothermic cracking reaction (kJ/kg feed), typically 350 kJ/kg feed (*see Table 3.4*)

Table 3.4 Some thermodynamic data for idealized reactions of importance in catalytic cracking[a] (from Venuto and Habib[1])

Reaction class	*Specific reaction*	*Log K_E (equlibrium constant)* 454°C	510°C	527°C	*Heat of reaction,* kJ/mole, 510°C
Cracking	n-$C_{10}H_{22} \rightarrow$ n-$C_7H_{16} + C_3H_6$	2.04	2.46	–	74.26
	1-$C_8H_{16} \rightarrow 2C_4H_8$	1.68	2.10	2.23	78.00
Hydrogen transfer	$4C_6H_{12} \rightarrow 3C_6H_{14} + C_6H_6$	12.44	11.09	–	–254.13
	cyclo-C_6H_{12} + 3 1-$C_5H_{10} \rightarrow$ 3n-$C_5H_{12} + C_6H_6$	11.22	10.35	–	–169.72
Isomerization	1-$C_4H_8 \rightarrow$ trans-2-C_4H_8	0.32	0.25	0.09[b]	–11.29
	n-$C_6H_{10} \rightarrow$ iso-C_4H_{10}	–0.20	–0.23	–0.36[b]	–7.92
	o-$C_6H_4(CH_3)_2 \rightarrow$ m-$C_6H_4(CH_3)_2$	0.33	0.30	–	–3.04
	cyclo-$C_6H_{12} \rightarrow CH_3$-cyclo-$C_5H_9$	1.00	1.09	1.10[b]	14.51
Transalkylation	C_6H_6 + m-$C_6H_4(CH_3)_2 \rightarrow 2C_6H_5CH_3$	0.65	0.65	0.65[b]	–0.51
Cyclization	1-$C_7H_{14} \rightarrow CH_3$-cyclo-$C_6H_{11}$	2.11	1.54	–	–88.00
Dealkylation	iso-C_3H_7-$C_6H_5 \rightarrow C_6H_6 + C_3H_6$	0.41	0.88	1.05[b]	94.07
Dehydrogenation	n-$C_6H_{14} \rightarrow$ 1-$C_6H_{12} + H_2$	–2.21	–1.52	–	129.77

[a]Data from *Technical Data Book, Petroleum Refining*, Division of Refining, American Petroleum Institute, New York (1966)

[b]Data from Voge[27]

Similarly for the regenerator:

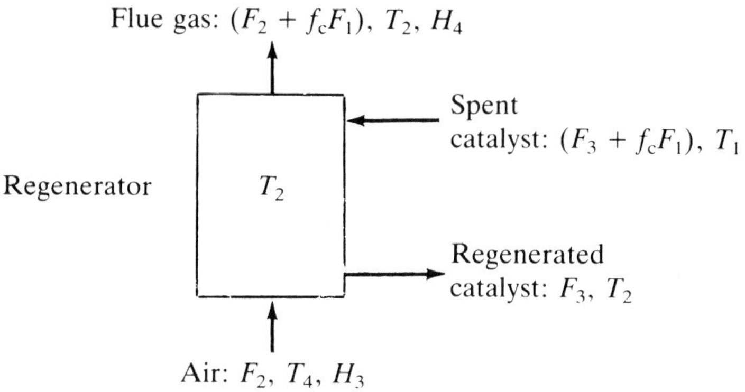

where F_2 = flowrate of air (kg/s)
T_4 = air temperature (°C)
H_3 = air enthalpy (kJ/kg)
H_4 = flue gas enthalpy (kJ/kg)

(heat released by coke combustion) = (heat gained by gases) + (heat gained by catalyst)

$$-\Delta H_{comb} f_c F_1 = [(F_2 + f_c F_1)H_4 - F_2 H_3] + F_3 C_{p,s}(T_2 - T_1) \qquad (3.4)$$

ΔH_{comb} is the heat of the exothermic coke combustion reaction (kJ/kg coke) and its value depends on a number of factors such as the carbon : hydrogen ratio in the solid (cited literature values are 7.6–11.0 per cent hydrogen[9]), and the extent of oxidation of C to CO_2. The heat of combustion of carbon to CO is considerably less than that of CO itself to CO_2:

	$\Delta H°$ kJ/mol[7]	
$C + \frac{1}{2}O_2 = CO$;	–121.3	
$CO + \frac{1}{2}O_2 = CO_2$;	–282.7	(3.5)
$C + O_2 = CO_2$;	–404.0	

CO : CO_2 ratios have in the past been typically of the order of unity but with the advent of controlled-combustion catalysts the complete oxidation of CO in the regenerator bed can now be achieved[1].

In equation 3.4 the small contribution of the coke deposit to the heat content of the catalyst is ignored.

A heat balance on the unit as a whole gives:

(heat released by coke combustion) = (heat of cracking reaction) + (heat gained by feed) + (heat gained by gases)

from which it follows that:

$$\frac{F_2}{F_1} = \frac{-\Delta H_{\mathrm{comb}} - \Delta H_{\mathrm{crack}} + [H_1 - (1 - f_c)H_2] - f_c H_4}{H_4 - H_3} \tag{3.6}$$

The air-to-oil ratio, F_2/F_1, is a function of both the coke content of the catalyst and the stoichiometry of the combustion reaction. Thus assuming the latter to be given by equation 3.5:

$$\frac{f_c F_1}{12} = \frac{0.21\ F_2}{1.293 \times 22.4}$$

i.e.

$$\frac{F_2}{F_1} = 11.493\ f_c \tag{3.7}$$

which would be increased by an appropriate amount if excess air were necessary to ensure complete combustion.

Since two of the primary variables in these equations, regenerator temperature and catalyst-to-oil ratio, F_3/F_1, are dependent variables, the heat balance calculation requires a trial-and-error approach. Thus for an assumed regenerator temperature, T_2, the catalyst coke content, f_c, may be found from equations 3.6 and 3.7 for given values of H_1, H_2, H_3, T_1, T_3 and T_4. With this value of f_c the catalyst-to-oil ratio, F_3/F_1, may be found using equation 3.3 and this may then be substituted into the regenerator heat balance equation 3.4, and the calculated regenerator temperature compared with the value assumed initially.

The highly coupled nature of the reactor-regenerator system is apparent when the results of changing operating variables from the steady state are considered. If, for example, the preheat temperature of the feed, T_3, is increased, the reactor temperature, T_1, will also rise causing the valve in the transfer line from the regenerator to be closed down thereby reducing the flow of hot catalyst to the reactor, i.e. F_3 is decreased. For a constant feedrate, F_1, the catalyst-to-oil ratio is thus reduced causing a fall in overall conversion, a corresponding reduction in coke yield and a fall in regenerator heat release. Heat balance is thus achieved at a slightly lower conversion, which can be restored to its previous level by allowing a slight rise in the temperature of the reactor. The effect of the lower coke rate in reducing regenerator temperature, T_2, is counteracted by the effect of the lower rate of removal of hot solids from the regenerator and in practice this causes T_2 to rise and $(T_2 - T_1)$ to increase. The coke level on the catalyst can be maintained at a lower value, since for a given overall conversion the coke yield decreases with increase in reactor temperature. Preheating of the feed thus allows more of it to be processed per hour for the same coke yield[37].

Table 3.5 Results of tests of Sohio's Toledo refinery (from Pohlenz[4])

Fresh feed rate, kg/s	44.4	44.4	44.4
Fresh feed temp., °C	212	298	370
Combined feed temp., °C	260	298	328
Combined feed ratio	2.3	2.3	2.3
Reactor temperature, °C	466	471	476
Regenerator temperature, °C	620	631	640
Reactor pressure, bar	2	2	2
Apparent conversion, vol %	69.7	69.7	69.7
Corrected conversion (196°C at 90% gasoline)	69.2	68.9	69.6
Catalyst circulation rate, kg/s	563	475	413
Product yield (% on feed):			
C_3 and lighter, wt %	7.6	7.8	9.1
Total C_4 fraction, wt %	8.8	8.6	7.7
Debutanized gasoline (196° F at 90%), vol %	49.6	50.1	51.2
Light cycle oil, vol %	26.1	27.6	26.7
Clarified slurry, vol %	4.2	3.5	3.7
Coke, wt %	9.4	8.6	7.9

Results of full-scale tests on the FCC unit at the Toledo refinery of Sohio have been reported[4] and are reproduced in *Table 3.5*. The purpose of these tests was to investigate the effect on the product spectrum of preheating the feed. As mentioned above, feed preheating causes a lower production of coke and consequently, if conversion is kept constant, a higher production of gasoline and other light products. In these tests the oil feed rate (28 000 b/d), combined feed ratio, conversion and catalyst content of the system were all held constant. The first two tests were conducted over a period of three days and the last over two days, mass balances in excess of 98.6 per cent being achieved in each case. The results show that as the feed-oil temperature was increased and the temperature of the reactor was raised to maintain a constant conversion, then:

(i) the coke yield fell in accordance with overall energy requirements
(ii) gasoline production increased correspondingly despite the possibility of overcracking at the higher reactor temperatures
(iii) the proportion of C_4 and lighter components remained constant although a change in distribution produced more C_2 and lighter material at the expense of C_3 and C_4.

The effect of the lower coke production on regenerator temperature is also apparent from the figures in *Table 3.5*. A more recent publication[8] reporting measurements made on a commercial riser unit also notes that a higher operating temperature results in a higher conversion and a net increase in gasoline yield.

3.6 Kinetics

One of the aims of conventional reaction kinetics is to predict the course of a chemical reaction by solving an integrated rate equation that contains measurable parameters such as reactant concentration. In the case of catalytic cracking reactions this procedure is made complicated by the problem of describing the system in precise kinetic terms. A typical cracking feedstock boiling in the range 220–540°C may contain tens of thousands of chemical compounds distributed among the various classes–alkanes (paraffins), cycloalkanes (naphthenes), aromatics, sulphur compounds, etc., all capable of reacting with the catalyst at different rates and to varying degrees. Under these circumstances the term 'reactant concentration' is, in the conventional sense, meaningless and it is necessary to lump species together in some way for kinetic analysis. One approach is to use the so-called 'n-d-M' method of Van Ness and Van Westen[38] in which refractive index, density and molecular weight are used to identify carbon atoms in the various structures. A more detailed procedure based on distillation, chromatography, mass spectroscopy and ultra violet spectroscopy has been reported by Fitzgerald *et al.*[39]. An application of the n-d-M method has been described by Rief *et al.*[40].

The situation is made more complicated by the coking reactions that not only cause the catalyst to lose activity more or less rapidly with time on stream but also bring about a change in its selectivity.

Nevertheless, it has been possible in recent years to develop kinetics suitable for incorporation into reactor–regenerator process models which can be used to predict the performance of commercial units[41].

3.6.1 Kinetics of cracking and coking

One of the earliest and most important studies of the kinetics of cracking and coking was that of Voorhies[42]. He determined the amount of coke deposited on both natural and synthetic clay catalysts at various temperatures and with a variety of charge stocks and found it to be a simple function of the catalyst residence time, t :

$$W_c = Pt^n \tag{3.8}$$

where W_c is the mass fraction of coke on the catalyst and P and n are constants, the value of n approximating to 0.5. Later studies of the cracking of pure compounds by Rudershausen and Watson[43] and by Prater and Lago[44] substantiated the relationship. An obvious difficulty with the

Voorhies equation however, and one that is discussed in his paper, is its implication that the rate of coke deposition is independent of reactant composition, extent of conversion and space velocity. Voorhies himself concluded this was due to a fortuitous balance between the coking tendencies of the feed and the products, but subsequent work has shown it to be the result of the investigation of too narrow a range of reaction variables (*see* Eberly *et al.*[45]). Nevertheless the Voorhies correlation has played a vital role in the design of commercial catalytic cracking units over the last thirty years[41].

Now the effect of the deposition of coke is clearly to reduce the reactivity of the catalyst towards the reactant in the cracking reaction, but the nature of this reactivity and the exact form in which it should be expressed in the kinetic equations has been the subject of some debate, and broadly speaking, two schools of though have emerged:

(i) Froment and Bischoff[46–48] base their analysis on a simple reaction mechanism $A \rightleftharpoons B$ with the conversion of A to B on a single site being the rate determining step:

$$\begin{aligned} A + l &\overset{K_A}{\rightleftharpoons} Al \\ Al &\overset{k_{sr}}{\longrightarrow} Bl \\ Bl &\overset{K_B}{\rightleftharpoons} B + l \end{aligned}$$

Here A and B represent the gas oil reactant and gasoline product respectively, l is a surface site, K_A and K_B are equilibrium constants and k_{sr} is a surface rate constant. Assuming a competetive reaction occurs leading to the formation of coke precursor Cl:

$$C + l \rightleftharpoons Cl$$

and the total concentration of active sites on the catalyst is:

$$C_t = C_l + C_{Al} + C_{Bl} + C_l \tag{3.9}$$

then a 'deactivation function' ϕ may be defined as the fraction of active sites remaining active on the cayalyst:

$$\phi = \frac{C_t - C_{Cl}}{C_t} \tag{3.10}$$

This function may be related to the catalyst coke content, C_c, by one or other of the following empirical relationships:

$$\phi = \exp(-\alpha C_c) \tag{3.11}$$

$$\phi = \frac{1}{1 + \alpha C_c} \tag{3.12}$$

where α is a constant. Thus the activity of the catalyst is defined in terms of its coke content:

$$\phi = f(C_c) \tag{3.13}$$

and the overall rate of the cracking reaction based on the above mechanism is:

$$-\frac{dC_A}{dt} = \frac{k_{sr}C_t K_A \phi \left(C_A - \dfrac{C_B}{K}\right)}{1 + K_A C_A + K_B C_B} \tag{3.14}$$

(ii) An alternative approach adopted by the Levenspiel[49,50] and Wojchiechowski[51] schools describes the cracking reaction in terms of the catalyst 'activity', a, defined as:

$$a = \frac{\text{rate at which catalyst converts reactant A}}{\text{rate of reaction of A with fresh catalyst}} \tag{3.15}$$

The activity itself is defined in terms of one of a number of empirical equations which, while similar to equations 3.11 and 3.12, differ in containing the catalyst residence time, t, rather than its coke content as the main variable. Two such equations are:

$$a = \exp(-k_d t) \tag{3.16}$$

$$a = \frac{k_d^{1/2}}{t^{1/2}} \tag{3.17}$$

where k_d is a deactivation rate constant. Equation 3.17 is derived directly from the Voorhies equation. Although equations 3.16 and 3.17 would appear to be quite different they were shown by Szepe and Levenspiel[49] to be special cases of the more general power law:

$$-\frac{da}{dt} = \kappa a^m \tag{3.18}$$

where κ is the deactivation rate constant and m the order of the deactivation reaction. Thus from equation 3.17 :

$$-\frac{da}{dt} = nP_2^{1/n}\, a^{(n+1)/n} \tag{3.19}$$

and so $\kappa = nP_2^{1/n}$ (3.20)

and $$m = \frac{n+1}{n} \tag{3.21}$$

n and P_2 being constants.
If the empirical value $n = 0.5$ is substituted in equation 3.21, it is apparent that the Voorhies equation corresponds to a third-order deactivation process:

$$-\frac{da}{dt} = k'_d a^3 \tag{3.22}$$

Differentiation of equation 3.16 leads to a first-order kinetic expression:

$$-\frac{da}{dt} = k_d a^3 \tag{3.23}$$

Kunii and Levenspiel[36] express the kinetics of the reactant conversion as a function of both catalyst activity and reactant concentration C_A :

$$-\frac{dC_A}{dt} = kC_A^r a^s \tag{3.24}$$

and give integrated forms of this equation for first-order cracking and decay kinetics (i.e. for $r = s = 1$), and for both plug and mixed flow of reactant:

$$\frac{C_A}{C_{Ao}} = \exp -\left(\frac{k\tau}{1 + k_d\bar{t}}\right) \quad \text{: plug flow} \tag{3.25}$$

$$\frac{C_A}{C_{Ao}} = \frac{1}{\left(1 + \dfrac{k\tau}{1 + k_d\bar{t}}\right)} \quad \text{: mixed flow} \tag{3.26}$$

where C_{Ao} and C_A are initial and final concentrations of reactant A, τ is the 'weight time' of the catalyst:

$$\tau = \frac{C_{Ao}W}{F_{Ao}} \tag{3.27}$$

W being the weight of catalyst in the reactant and F_{Ao} the reactant flowrate; $\bar{t}$ is the mean residence time of the catalyst circulating with velocity F_s :

$$\bar{t} = \frac{W}{F_s} \tag{3.28}$$

For the more complex kinetics of Froment and Bischoff given earlier the cracking rate in terms of a would be:

$$-\frac{dC_A}{dt} = \frac{k_{sr}K_A e^{-k_d t}\left(C_A - \dfrac{C_B}{K}\right)}{1 + K_A C_A + K_B C_B} \tag{3.29}$$

Comparing the two approaches (*see* Froment and Bischoff[48]), it is clear that the second one is the easier to apply since it contains the process time as a main variable, whereas the first method requires an additional equation for the rate of coke formation in order to introduce the time variable. However the parameter k_d is not, unlike the α of equations 3.11 and 3.12, a fundamental quantity but a function of the operating variables and must vary as the operating conditions change. Furthermore, the Levenspiel approach by omitting a coke concentration term from the deactivation rate equation appears to predict a uniform rate of coking throughout the catalyst whereas coke profiles which are known to develop within particles lead to non-uniform deactivation. Despite these objections, process models based on the time-dependent deactivation rate equations have been developed and used successfully in the control and optimization of catalytic cracking operations.

3.6.2 Regeneration

Coked catalyst from the cracking reactor is fed to the regenerator unit where it is fluidized with air and the coke burned off; hot, regenerated catalyst is then returned to the cracking reactor. The efficiency of the regenerator is thus determined by the rate of coke combustion and by the extent of its conversion to CO_2. Ideally the oxygen level in the flue gas should be in the range 0–0.2 per cent[1] thereby ensuring a low level of carbon on the regenerated catalyst and minimizing the extent of afterburning in the freeboard region. The latter can cause uncontrolled temperature excursions above the bed (temperatures of over 780°C have been reported[52]), leading to serious damage of both catalyst and cyclones. The modern use of controlled combustion catalysts has, however, greatly reduced the risk of afterburning and the emission of CO from the regenerator can now be practically eliminated.

Kinetics of coke burning on cracking catalysts of the spherical Durabead type (d_p = 2–5 mm) were studied by Weisz and Goodwin[53–55] and were shown to be of the form:

$$-\frac{dC_c}{dt} = k_2 C_A C_c \tag{3.30}$$

where C_c is the concentration of coke on the catalyst, C_A the concentration of oxygen at the gas–coke interface and k_2 is a second-order rate constant. For 2 mm diameter beads at temperatures above 450°C the combustion mechanism was found to be of the 'shrinking core' or 'shell progressive' type, with the kinetics being controlled by the rate of pore diffusion of oxygen to the coke surface. Butt has reviewed this mechanism in the case of fixed-bed reactors[56]. For finely powdered material ($d_p \simeq$ 0.1 mm) however, Weisz and Goodwin found that reaction occured uniformly throughout a particle over the complete temperature range studied (450–600°C) and that the combustion kinetics were chemically controlled (*Figure 3.9*). These 'intrinsic' kinetics were found to be first order in carbon concentration and to have a rate constant in air of:

$$k_1 = 4 \times 10^7 \exp[-(36000 \pm 1600)/RT] \tag{3.31}$$

where the activation energy is given in cal/mol. This value was shown to be closely similar to the value obtained by other workers for the combustion of graphite. Weisz and Goodwin also showed that the kinetics are

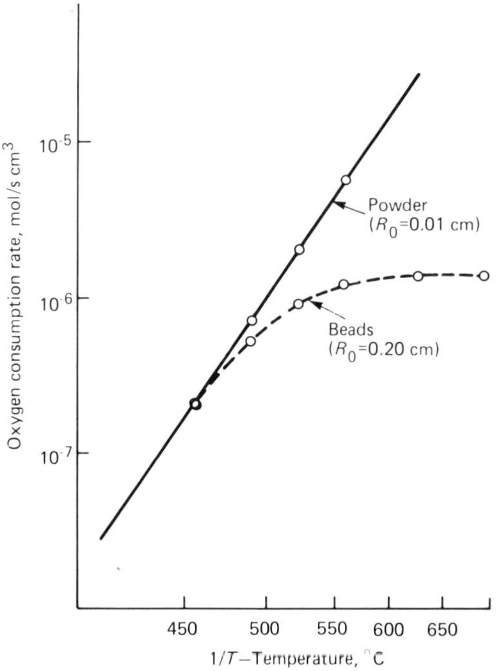

Figure 3.9 Burning rates of coke on Durabead cracking catalyst (From Weisz and Goodwin[53])

independent of the source and structure of the deposited coke but that the rate of burn-off is greatly accelerated by the presence on the surface of transition metal oxides.

A more recent study by Hano *et al.*[57] of the combustion of coke on zeolites has produced results that are largely in agreement with those of Weisz and Goodwin. They showed that in the region of chemical control the kinetics were first order in both coke concentration and oxygen partial pressure, although both the pre-exponential factor (1.8×10^7) and activation energy (26.2 kcal/mol) were somewhat lower than in the case of silica–alumina. Hano *et al.* also showed that the burning rate was independent of the source of coke but was accelerated by the presence of certain cations in the zeolite structure. Thus the rate constant for a copper exchanged Y-zeolite (Linde SK–40) was fourteen times greater than that for the same material exchanged with cadmium.

3.7 Process models

3.7.1 Cracking and coking

Recently Weekman, Nace and co-workers of the Mobil company[41,59–62] have developed a process model in which the complex reactant-product mixture is lumped into a three-component system reacting in the following way:

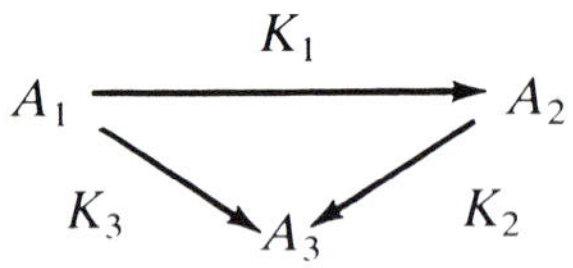

Here A_1 represents the mass fraction of unreacted gas oil, A_2 the mass fraction of gasoline product and A_3 the mass fraction of C_1 to C_4 gases and coke. The K terms are rate constants. Now although the cracking of pure hydrocarbons is known to follow first-order kinetics[28,31,60,63,64], the Mobil model assumes the cracking of the gas oil feed to be a second-order process. This is thought to be reasonable in view of the wide range of cracking rates of the individual components–the heavier constituents cracking more slowly than the lighter ones and the overall process being well represented by a second order kinetic scheme. Theoretical justification for this lumping procedure and the kinetic formulation has been provided by the work of Luss *et al.*[65–67]. Because of its narrower spectrum of molecular weights the gasoline product is considered to crack according to first-order kinetics. The differential equations describing the system for a vapour-phase, plug-flow, isothermal reactor (whether fixed, fluid or moving bed) are therefore:

$$\frac{dA_1}{dt_0} = -(K_1 + K_3)A_1^2 a \tag{3.32}$$

$$\frac{dA_2}{dt_0} = K_1A_1^2a - K_2A_2a \tag{3.33}$$

$$A_3 = 1 - A_1 - A_2 \tag{3.34}$$

Here t_0 is the oil residence time (s) and a is the catalyst activity (Section 3.6.1). Now if the ratio of the volumetric flowrate of catalyst to the volumetric flowrate of oil is R, then:

$$R = \frac{t_0}{t_c}$$

t_c being the catalyst residence time (s). Hence, substituting for t_0 in equation 3.32, putting $K_0 = K_1 + K_2$, and assuming the catalyst decay function to be represented by:

$$a = a_0 \exp(-k_d t_c)$$

where a_0 is the activity of the non-fully regenerated catalyst entering the reactor:

$$\frac{dA_1}{R dt_c} = -K_0A_1^2\, a_0 \exp(-k_d t_c) \tag{3.35}$$

which may be integrated with boundary conditions

$$A_1 = 1 \text{ at } t_c = 0$$

$$A_1 = A_1 \text{ at } t_c = t_c$$

to give:

$$A_1 = \frac{1}{1 + \dfrac{K_0a_0R}{k_d}\left[1 - \exp(-k_d t_c)\right]} \tag{3.36}$$

If S is the space velocity (volume of oil/total volume of reactor unit time) then R may be replaced by $1/St_c$, giving for the overall oil conversion:

$$\varepsilon = 1 - A_1 \tag{3.37}$$

$$= \frac{\dfrac{K_0a_0}{S}\left[1 - \exp(-k_d t_c)\right]}{k_d t_c + \dfrac{K_0a_0}{S}\left[1 - \exp(-k_d t_c)\right]} \tag{3.38}$$

The expression for the production rate of gasoline is somewhat more complicated, due to the combination of second-order formation and first-order reaction terms (equation 3.33).

Hence:

$$A_2 = k_1 k_2 \exp(-k_2/A_1)\left[\frac{\exp(k_2)}{k_2} - \frac{A_1}{k_2}\exp(k_2/A_1) - E(k_2) + E\left(\frac{k_2}{A_1}\right)\right] \quad (3.39)$$

$$\text{where } k_1 = \frac{K_1}{K_0}\ ;\ k_2 = \frac{K_2}{K_0}$$

$$E(x) = \int_{-\infty}^{x} \frac{e^x}{x}\,dx \quad \text{(the exponential integral)}$$

As already mentioned, these solutions apply to any isothermal reactor in which both oil vapour and catalyst are in plug flow. Weekman and Nace also give solutions for a fluid bed reactor where the catalyst is assumed to be perfectly mixed, the gas still being in plug flow:

$$\varepsilon\ \text{(fluid bed)} = \frac{\dfrac{K_0}{S}}{\left[1 + k_d t_c + \left(\dfrac{K_0}{S}\right)\right]} \quad (3.40)$$

and for a fixed-bed reactor with plug flow of gas:

$$\varepsilon\ \textit{(fixed bed)} = \frac{\left(\dfrac{K_0}{S}\right)\exp(-k_d t_c)}{1 + \left(\dfrac{K_0}{S}\right)\exp(-k_d t_c)} \quad \textit{(instantaneous)} \quad (3.41)$$

$$\bar{\varepsilon}\ \text{(fixed bed)} = \frac{1}{k_d t_c}\, ln\left[\frac{1 + \left(\dfrac{K_0}{S}\right)}{1 + \left(\dfrac{K_0}{S}\right)\exp(-k_d t_c)}\right] \quad \text{(time averaged)} \quad (3.42)$$

These equations were tested experimentally by Nace *et al.* for a number of different charge stocks[61]. The time-averaged conversion (equation 3.42) plotted against space velocity for a highly paraffinic charge stock of wide boiling range is shown in *Figure 3.10*. The agreement between experimental points and the computed curves is clearly very good. Similar agreement is shown for the gasoline selectivity data in *Figure 3.11*, from

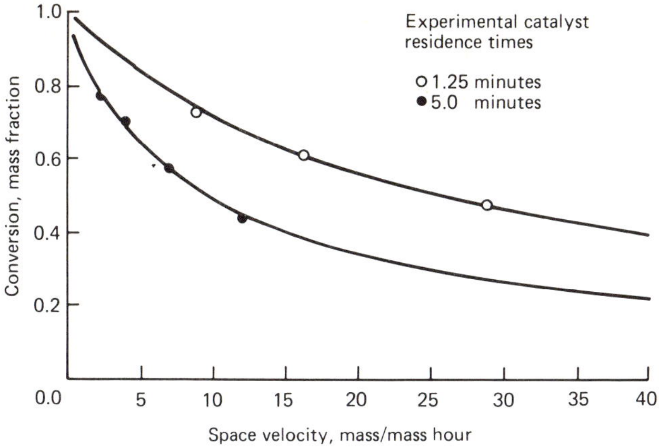

Figure 3.10 Conversion v. space velocity (From Nace et al.[61])

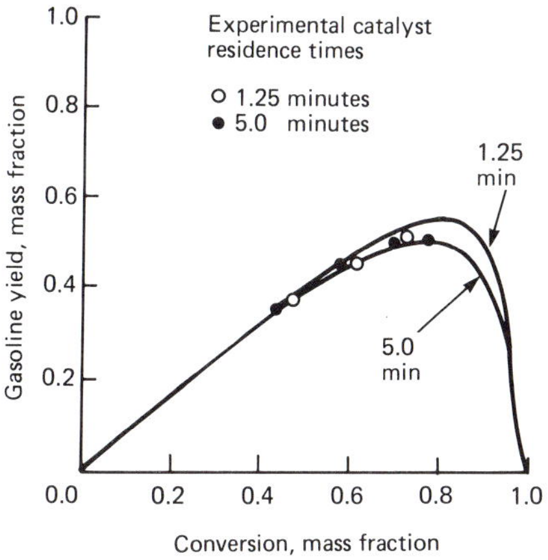

Figure 3.11 Gasoline selectivity (From Nace et al.[61])

which also the effect of overcracking at high conversions is apparent. This experimental work showed the feedstocks with high concentrations of paraffins and/or naphthenes to have the highest rate constants for gas oil cracking and gasoline formation; recycle and coker feeds were shown to have lower rate constants than virgin oils. In other words, the zeolite catalysts were shown to crack paraffins and naphthenes rapidly and with excellent selectivity, but to crack aromatics much more slowly with poor selectivity. Furthermore high concentrations of aromatics in the feed were the cause of increased rates of deactivation.

Despite the success of the Weekman–Nace model in predicting conversions from the range of charge stocks investigated, it was not found possible to extrapolate to other charge stocks outside the range of the correlation.

Consequently a more complex lumping scheme was developed by Jacob *et al.*[41] containing ten lumped species that interact according to the scheme shown in *Figure 3.12*.

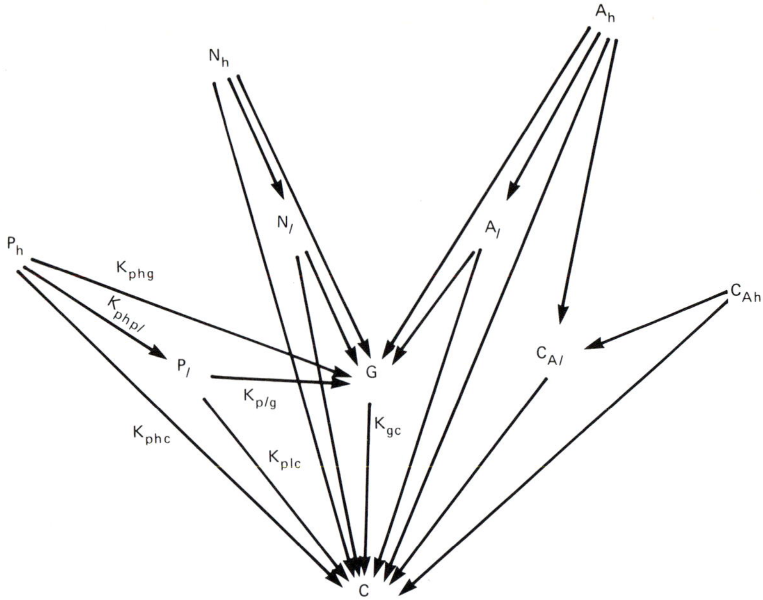

P_l = Wt. % paraffinic molecules, (mass spec analysis), 220–340°C
N_l = Wt. % naphthenic molecules, (mass spec analysis), 220–340°C
C_{Al} = Wt. % carbon atoms among aromatic rings, (n-d-M method),220–340°C
A_l = Wt. % aromatic substituent groups (220–340°C)
P_h = Wt. % paraffinic molecules, (mass spec analysis), >340°C
N_h = Wt. % naphthenic molecules, (mass spec analysis), >340°C
C_{Ah} = Wt. % carbon atome among aromatic rings, n-d-M method, >340°C
A_h = Wt. % aromatic substituent groups (>340°C)
G = G lump (C_5–220°C)
C = C lump (C_1 to C_4 + COKE)
$C_{Al} + P_l + N_l + A_l$ = LFO(220–340°C)
$C_{Ah} + P_h + N_h + A_h$ = HFO(>340°C)

Adapted Nomenclature for rate constants is detailed in the above Figure for the paraffinic molecules. Similar rules apply for the other reaction steps.

Figure 3.12 Lumped kinetic scheme (Jacob et al.[41])

The resulting rate constants, determined using a laboratory-scale fixed-fluidized dense bed reactor, are invariant with respect to charge stock composition.

A similar approach to that of the original Weekman–Nace theory has also been adopted by Paraskos *et al.* of Gulf[68,69] for modelling catalytic cracking in a transfer line (i.e. riser) reactor. Their model is based on a three lump kinetic scheme with two rate constants:

$$
\begin{array}{lll}
 & k_1 & \\
A_1 & \rightarrow A_2 & + \; A_3 \\
 & \downarrow k_2 & \\
 & A_3 &
\end{array}
$$

the symbols having the same significance as before. Catalyst deactivation is assumed to be of the form:

$$a = \exp(-k_d\eta) \tag{3.43}$$

where η, the so-called 'flowing space time', is defined as:

$$\eta = \frac{F_c}{F_f} \times \frac{1}{A_1} \times \bar{t}, \tag{3.44}$$

F_c and F_f being the catalyst and feed mass flow rates (kg/s) and $\bar{t}$ the contact time (s):

$$\bar{t} = \frac{V_R}{\rho F_f} \tag{3.45}$$

V_R is the reactor volume and ρ the density of the feed vapour at the inlet. Gas oil conversion is then found from the mass balance equation:

$$\frac{dA_1}{d\eta} = -k_1^0 A_1^2 \exp(-k_d\eta) \tag{3.46}$$

assuming second order kinetics, k_1^0 being the rate constant for gas oil cracking at $\eta = 0$.

From equation 3.46 the authors obtain:

$$\frac{1}{A_1} - 1 = \frac{k_1^0}{k_d}[1 - \exp(-k_d\eta)] \tag{3.47}$$

$$\text{i.e. } A_1 = \frac{1}{1 + \dfrac{k_1^0}{k_d}[1 - \exp(-k_d\eta)]} \tag{3.48}$$

which is closely similar to equation 3.36. Pilot plant data obtained with a transfer line reactor operated isothermally are shown in *Figures 3.13* and *3.14*, the group η being used to correlate the results. Clearly at low values of space time gasoline yields increase with temperature, but at higher values the effect of temperature diminishes considerably. Similar trends were found for the light gases and coke.

The foregoing model has been extended by Shah *et al.*[69] to the case of adiabatic operation of the reactor. Here the temperature of the reactant

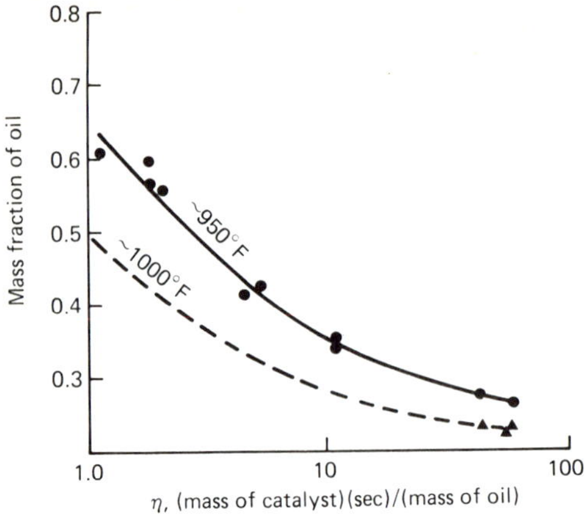

Figure 3.13 Temperature–flowing space time effects in gas oil cracking (From Paraskos et al.[68])

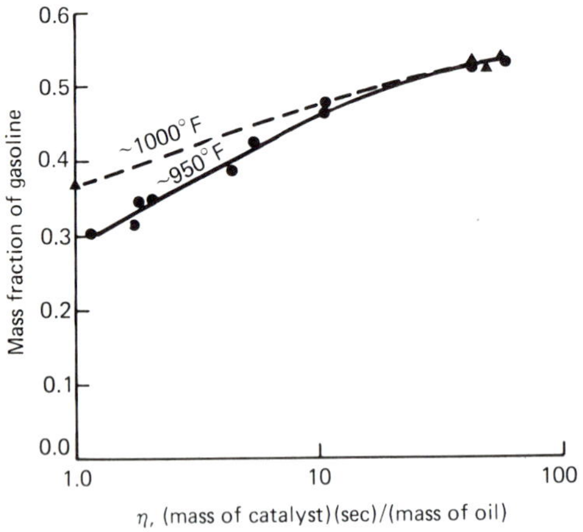

Figure 3.14 Temperature–flowing space time effects in gasoline production (From Paraskos et al.[68])

mixture decreases with increasing contact time thereby causing the cracking reaction to slow down. Assuming catalyst activity to be independent of temperature the steady state mass balance of gas oil becomes:

$$\frac{dA_1}{d\eta} = -k_1^0 A_1^2 \exp[-\delta/(1+\theta)] \exp(-k_d\eta) \qquad (3.49)$$

where $\delta = \dfrac{E}{RT_i}$

and $\theta = \dfrac{T - T_i}{T_i}$

E is the activation energy of the gas oil cracking reaction, T_i the fluid inlet temperature and T the temperature at any point along the reactor. Shah *et al.* give a series solution for equation 3.49 and find its predictions to be in good agreement with experimental data from a commercial adiabatic riser reactor.

The use of process models such as those described here is increasing in industrial operations, although relatively few reports concerning them appear in the open literature. They are extremely valuable for optimization and control particularly of processes as complex as fluidized catalytic cracking.

3.7.2 Regenerator models

The regenerator of a catalytic cracking system is a fluidized bed reactor in which an essentially non-catalytic gas–solid reaction, the combustion of catalyst coke, is carried out. The performance of the reactor is measured in terms of the efficiency of coke removal and the level to which it can be reduced on the regenerated catalyst, and of the extent of conversion of CO to CO_2. At the temperatures in question (600–700°C) both reaction kinetics and mass transfer are likely to be of importance, and given the necessary parameters, it should be possible to use one of the reactor models discussed in Chapter 2 to predict the performance of the unit, particularly in view of the much simpler chemistry of the process compared with that of the cracking reaction itself.

An early attempt to model a regenerator was described by Pansing[11] in 1956. He proposed a simple model based on the kinetic scheme of equation 3.30 coupled with the diffusion of oxygen from gas bubbles to the emulsion phase represented by:

$$r_{\text{diff}} = k_{\text{diff}}\,(C_{\text{A}}\text{s2b} - C_{\text{A}}\text{s2e}) \qquad (3.50)$$

where r_{diff} is the rate of oxygen diffusion, k_{diff} is a mass transfer coefficient and $C_{\text{A}}\text{s2b}$ and $C_{\text{A}}\text{s2e}$ are the concentrations of oxygen in the bubble and emulsion phases respectively. Under steady-state conditions the overall rate of reaction is:

$$r = \frac{C_{\text{A}}\text{s2b}}{\dfrac{1}{k_{\text{diff}}} + \dfrac{1}{k_2 C_{\text{c}}}} \qquad (3.51)$$

The fraction of oxygen unconverted at the surface of the bed may be obtained by writing a mass balance on oxygen across a differential length of the unit:

$$F_{\text{o}}\text{d}C_{\text{A}} = -r\text{d}M \qquad (3.52)$$

where F_{o} is the gas feed rate and M is the total mass of catalyst. Then assuming plug flow of gas and a uniform distribution of coke throughout the emulsion, equations 3.51 and 3.52 may be combined and integrated to give:

$$\frac{C_A}{C_{Ao}} = \exp - \left[\frac{M/F_o}{(1/k_{diff} + 1/k_2 C_c)} \right] \tag{3.53}$$

Pansing determined the kinetic and mass transfer coefficients in euqation 3.53 using a 10 in diameter pilot-scale reactor, but found poor agreement between the resulting predictions of the model and the performance of a 30 ft diameter commercial unit.

More recently, a description of regenerator performance has been given by Kunii and Levenspiel[36] in terms of their 'bubbling bed' model of fluid bed reactors (Section 2.3.4). For kinetics of the Weisz–Goodwin type that are first order in reactant gas, and assuming no restriction on continuous reaction throughout a particle, the concentration of coke on catalyst is shown to be:

$$\frac{C_c}{C_{co}} = \int_{t=0}^{\infty} \exp(-k_r \bar{C}_{Ae} t) \exp(-t/\bar{t}) \, dt \tag{3.54}$$

$$= \frac{1}{1 + k_r \bar{C}_{Ae} \bar{t}} \tag{3.55}$$

Where $\bar{C}_{Ae}$ is the average concentration of oxygen in the emulsion phase, k_r is a rate coefficient based on unit volume of solid, and $\bar{t}$ is the mean residence time of particles in the bed, i.e.:

$$\bar{t} = \frac{M}{F_s} \tag{3.56}$$

where F_s is the flowrate of catalyst out of the system and M is the mass in the regenerator.

From the bubbling bed model $\bar{C}_{Ae}$ is evaluated as follows.

$$\bar{C}_{Ae} = \frac{C_{Ao}[1 - \exp(-k_f)]}{(H_f k_r / U_b)} \phi \tag{3.57}$$

where k_f is the global rate coefficient as defined in equation 2.65 of Section 2.3.4, H_f is the bed height, U_b is the rise velocity of a gas bubble in isolation:

$$U_b = 0.711(g d_b)^{1/2} \text{ (Chapter 1, Section 1.6.3)}$$

and ϕ is defined as:

$$\phi = \left[\frac{1}{1 + \gamma_e \dfrac{k_r}{K_{ce}}}\right] \left[\frac{1}{\gamma_c + \dfrac{1}{\dfrac{k_r}{K_{ce}} + \dfrac{1}{\gamma_e}}}\right] \tag{3.58}$$

In this expression γ_c and γ_e are the ratios of solids in the cloud–wake and emulsion phases respectively to the total volume of bubbles in the bed, while K_{ce} is a cloud-to-emulsion gas transfer parameter (see Section 2.2.2.1). The concentration of oxygen in the exit gas is then found from a material balance on the solid and gas, which for an overall stoichiometry of $C + O_2 = CO_2$ is:

$$\frac{F_s}{12}\left(1 - \frac{C_c}{C_{co}}\right) = F_o C_{A,o}\left(1 - \frac{C_A}{C_{Ao}}\right) \tag{3.59}$$

where F_s and F_o are the flowrates of solid and air respectively (kg/s). The Kunii–Levenspiel model has been applied to an industrial unit by de Lasa and his co-workers and found in all cases to predict too low a conversion due, it would seem, to its underestimate of the extent of oxygen transfer from bubbles to the emulsion phase. This work will now be considered in more detail.

An extra degree of complexity has been introduced into regenerator modelling by the studies of de Lasa and his co-workers who have considered the effects on the unit of reaction in the grid and freeboard regions. An interesting aspect of the work has been their ability to compare the predictions of theoretical models with the operating performance of the industrial regenerator unit of Destileria La Plata, YPF, Argentina; the unit was 10 m high, 14 m in diameter and contained some 270 tonnes of catalyst[70]. Reaction in the grid region coupled with the burning of coke in the bed itself was considered by Errazu *et al.*[71] who developed a model based on the following assumptions:

(i) Fluidizing air enters the bed as jets which penetrate to a small finite distance in the bed before breaking up into bubbles. Jets and bubbles are of uniform size.
(ii) Jets are in plug flow while bubbles are completely mixed.
(iii) Combustion occurs only in the emulsion phase of the fluid bed which behaves as a completely stirred tank reactor (CSTR).
(iv) Coke burning rates are given by the kinetic equations of Weisz and Goodwin[53–55] and the CO/CO_2 ratio by the function proposed by Arthur[72].
(v) The coke may be represented by the formula CH_2 and burns according to:

$$CH_2 + O_2 \rightarrow CO + H_2O$$
$$CH_2 + 1\tfrac{1}{2}O_2 \rightarrow CO_2 + H_2O$$

(vi) Homogeneous oxidation of CO to CO_2 does not occur at the temperatures considered, i.e. below 670°C.

Mass balance equations on oxygen and coke are then set up, solved analytically and combined with an adiabatic heat balance equation to give the main operating variables in terms of the model parameters. These parameters relate to the fluid dynamic, mass transfer and kinetic processes of the system and may be varied at will. The authors used the correlations of Basov *et al.*[73] to compute jet penetration and bubble size, and the measurements of Behie *et al.*[74,75] to determine mass transfer coefficients for jets formed at the grid holes; interphase mass transfer in the bed itself was estimated using the correlations of Davidson and Harrison[76] and Kunii and Levenspiel[36]. The model equations were solved for operating conditions similar to those used on the La Plata regenerator and it was concluded that the bed could be well simulated with the CSTR model and that the grid region exerted little influence on performance.

The ineffectiveness of the grid region is perhaps surprising in view of the observations of Rowe *et al.*[77] on the discharge of gas from an orifice into a fluidized bed (Chapter 1, Section 1.14 and Chapter 2, Section 2.5.1). The influence of the freeboard region on regenerator performance has however been shown to be significant. De Lasa and Grace[71] used the above mentioned model of Errazu *et al.* for the grid and bed regions but added a term for particles thrown up above the bed surface; freeboard trajectories were calculated by the method of George and Grace[78] (*see* Section 1.9). The resulting model predictions are shown in *Figure 3.15*, where the effect

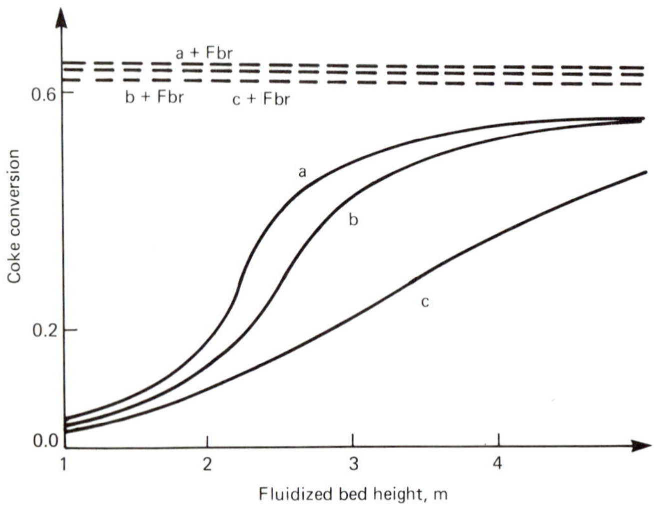

Figure 3.15 Influence of the model chosen to represent the dense bed part of the fluidized bed, with and without the freeboard region considered (From de Lasa and Grace[70]). (a) CSTR model; (b) Davidson and Harrison model; (c) as (b), but with lower interphase mass transfer given by Kunii and Levenspiel: $(H + H_F) = 10$ m, $Q = 73.9$ m^3/s, $F_s = 506$ kg/s, $T^0 = 474$°C

on coke conversion of including a freeboard reaction term in three different fluid bed models is compared. It is clear that freeboard effects are expected to be significant but to be relatively unaffected by the choice of fluid bed model. However, from a comparison between the overall model predictions and actual operating data from the La Plata regenerator, de Lasa *et al.*[79] found that the discrepancy in the coke conversion caused by not including a freeboard term was only some 2 to 4 per cent and they concluded that the complexity introduced by the combined three-region model was not justified in the case of deep regenerators (3.5–4 m deep). In all cases a CSTR model proved to be quite adequate.

References

1 VENUTO, P.B. and HABIB, T.E., *Fluid Catalytic Cracking with Zeolite Catalysts*, Marcel Dekker Inc., New York (1979)

2 JAHNIG, C.E., CAMPBELL, D.L. and MARTIN, H.Z., in *Fluidization*, Eds Grace, J.R. and Matsen, J.M., 3, Plenum Press, New York (1980)

3 TURNER, R., *Soc. Chem. Ind. Symp. Fluidization*, London, 47 (1964)

4 POHLENZ, J.B., *Oil Gas J.*, **61**, 124 (1963)

5 REHBEIN, C.A., MITCHELL, W.A. and WILSON, R.A., *Oil Gas J.*, **25**, 108 (1959)

6 WEISZ, P.B., *Chemtech*, 498 (1973)

7 HOUGEN, O.A., WATSON, K.M. and RAGATZ, R.A., *Chemical Process Principles*, 2nd Edn., Part 1, 502, Wiley, New York (1954)

8 SCHUURMANS, H.J.A., *Ind. Eng. Chem. Proc. Des. Dev.*, **19**, 267 (1980)

9 WHITTINGTON, E.L., MURPHY, J.R. and LUTZ, I.H., *ACS, Div. Petrol. Chem.*, **17**(3), B-66 (1972)

10 SAXTON, A.L. and WORLEY, A.C., *Oil Gas J.*, **68**, 82 (1970)

11 PANSING, W.F., *A.I.Ch.E. Jl*, **2**, 71 (1956)

12 RYLAND, L.B., TAMELE, M.W. and WILSON, J.N., in *Catalysis*, Ed. Emmett, P.H., **6**, 1 Reinhold, New York (1956)

13 WELLS, A.F., *Structural Inorganic Chemistry*, Oxford University Press, 4th Edn (1975)

14 BRECK, D.W., EVERSOLE, W.G. and MILTON, R.M., *J. Amer. Chem. Soc.*, **78**, 2338 (1956)

15 BRECK, D.W. and FLANIGEN, E.M., *Symp. Molecular Sieves, Soc. Chem. Ind.*, London, 47 (1968)

16 MILTON, R.M., US Patent 2882243 (1959)

17 BRECK, D.W., EVERSOLE, W.G., MILTON, R.M., REED, T.B. and THOMAS, T.L., *J. Amer. Chem. Soc.*, **78**, 5963 (1956)

18 MILTON, R.M., US Patent 2882244 (1959)

19 MILTON, R.M., US Patent 3030181 (1962)

20 MIALE, J.C., CHEN, N.Y. and WEISZ, P.B., *J. Catalysis*, **6**, 278 (1966)

21 GATES, B.G., KATZER, J.R. and SCHUIT, G.C.A., *Chemistry of Catalytic Processes*, McGraw-Hill, New York (1979)

22 VAUGHN, D.E.W., in *Properties and Applications of Zeolites*, Ed. Townsend, R.P., Special Publication No. 33, Chem. Soc., London (1980)

23 LETZSCH, W.S., RITTER, R.E. and VAUGHN, D.E.W., *Oil Gas J.*, **69**, 130 (1971)

24 BLAZEK, J.J., *Oil Gas J.*, **69**, 66 (1971)

25 APPLEBY, W.G., GIBSON, J.W. and GOOD, G.M., *Ind. Eng. Chem. Proc. Des. Dev.*, **1**, 102 (1962)

26 GREENSFELDER, B.S., VOGE, H.H. and GOOD, G.M., *Ind. Engng Chem.*, **41**, 2573 (1949)

27 VOGE, H.H., in *Catalysis*, Ed. Emmett, P.H., **6**, 407, Reinhold, New York (1958)

28 HAENSEL, V., Adv. in *Catalysis*, **3**, 179 (1951)

29 MACKOR, E.L., HOFSTRA, A. and VAN DER WAALS, J.H., *Trans. Faraday Soc.*, **54**, 186 (1958)

30 OZAWA, Y. and BISCHOFF, K.B., *Ind. Eng. Chem. Proc. Des. Dev.*, **7**, 78 (1968)

31 SHEPHARD, F.E., ROONEY, J.J. and KEMBALL, C., *J. Catalysis*, **1**, 379 (1962)
32 VENUTO, P.B., in *Catalysis in Organic Synthesis*, Ed. Smith, G.V., Academic Press, New York (1977)
33 JOHN, T.M. and WOJCHIECHOWSKI, B.W., *J. Catalysis*, **37**, 240 (1975)
34 JOHN, T.M., PACHOVSKY, R.A. and WOJCHIECHOWSKI, B.W., *Chem. Reaction Engng*, **2**, 422 (1974)
35 VENUTO, P.B., HAMILTON, L.A. and LANDIS, P.S., *J. Catalysis*, **5**, 484 (1966)
36 KUNII, D. and LEVENSPIEL, O., *Fluidization Engineering*, Wiley, New York (1969)
37 HOBSON, G.D. and POHL, W., *Modern Petroleum Technology*, 4th Edn, 305, Institute of Petroleum, London (1973)
38 VAN NEES, K. and VAN WESTEN, H.A., *Aspects of the Constitution of Mineral Oils*, Elsevier, New York (1951)
39 FITZGERALD, M.E., MOIRANO, J.L., MORGAN, H. and CIRILLO, V.A., *Appl. Spectrosc.*, **24**, 106 (1970)
40 RIEF, H.E., KRESS, R.F. and SMITH, J.S., *Petrol. Refiner*, **40**, 237 (1961)
41 JACOB, S.M., GROSS, B., VOLTZ, S.E. and WEEKMAN, V.W., *A.I.Ch.E. Jl*, **22**, 701 (1976)
42 VOORHIES, A., *Ind. Engng Chem.*, **37**, 318 (1945)
43 RUDERSHAUSEN, C.G. and WATSON, C.C., *Chem. Engng Sci.*, **3**, 110 (1954)
44 PRATER, C.D. and LAGO, R.M., Adv. in *Catalysis*, **8**, 294 (1956)
45 EBERLY, P.E., KIMBERLIN, C.N., MILLER, W.H. and DRUSHEL, H.V., *Ind. Engng Chem. Proc. Des. Dev.*, **5**, 193 (1966)
46 FROMENT, G.F. and BISCHOFF, K.B., *Chem. Engng Sci.*, **16**, 189 (1961)
47 FROMENT, G.F. and BISCHOFF, K.B., *Chem. Engng Sci.*, **17**, 105 (1962)
48 FROMENT, G.F. and BISCHOFF, K.B., *Chemical Reactor Analysis and Design*, Wiley, New York (1979)
49 SZEPE, S. and LEVENSPIEL, O., *Proc. 4th Eur. Symp. Reaction Engng*, Brussels (1968), 265, Pergamon Press, Oxford (1971)
50 LEVENSPIEL, O., *Chemical Reaction Engineering*, 2nd Edn, Wiley, New York (1972)
51 WOJCHIECHOWSKI, B.W., *Can. J. Chem. Engng*, **46**, 48 (1968)
52 MURPHY, J.R. and SOUDEC, M., *Oil Gas J.*, **75**, 70 (1977)
53 WEISZ, P.B. and GOODWIN, R.D., *J. Catalysis*, **2**, 397 (1963)
54 WEISZ, P.B. and GOODWIN, R.B., *J. Catalysis*, **6**, 227 (1966)
55 WEISZ, P.B., *J. Catalysis*, **6**, 425 (1966)
56 BUTT, J.B., Adv. in *Chem. Ser.*, **109**, 259 (1972)
57 HANO, T., NAKASHIO, F. and KUSONOKI, K., *J. Chem. Eng. Japan*, **8**, 127 (1975)
58 WEEKMAN, V.W., *Ind. Eng. Chem. Proc. Des. Dev.*, **7**, 90 (1968)
59 NACE, D.M., *Ind. Eng. Chem. Prod. Res. Dev.*, **7**, 203 (1970)
60 WEEKMAN, V.W. and NACE, D.M., *A.I.Ch.E. Jl*, **16**, 397 (1970)
61 NACE, D.M., VOLTZ, S.E. and WEEKMAN, V.W., *Ind. Eng. Chem. Proc. Des. Dev.*, **10**, 530 (1971)
62 VOLTZ, S.E., NACE, D.M., JACOB, S.M. and WEEKMAN, V.W., *Ind. Engng Chem. Proc. Des. Dev.*, **11**, 261 (1972)
63 BLANDING, F.H., *Ind. Engng Chem.*, **45**, 1186 (1953)
64 SATTERFIELD, C.N., *A.I.Ch.E. Jl*, **21**, 209 (1975)
65 HUTCHINSON, P. and LUSS, D., *Chem. Engng J.*, **1**, 129 (1970)
66 LUSS, D. and HUTCHINSON, P., *Chem. Engng J.*, **2**, 172 (1971)
67 GOLIKERI, S. and LUSS, D., *A.I.Ch.E. Jl*, **18**, 277 (1972)
68 PARASKOS, J.A., SHAH, Y.T., McKINNEY, J.D. and CARR, N.L., *Ind. Engng Chem. Proc. Des. Dev.*, **15**, 165 (1976)
69 SHAH, Y.T., HULING, G.P., PARASKOS, J.A. and McKINNEY, J.D., *Ind. Engng Chem. Proc. Des. Dev.*, **16**, 89 (1977)
70 DE LASA, H.I. and GRACE, J.R., *A.I.Ch.E. Jl*, **25**, 984 (1979)
71 ERRAZU, A.F., DE LASA, H.I. and SARTI, F., *Can. J. Chem. Engng*, **57**, 191 (1979)
72 ARTHUR, J.R., *Trans. Faraday Soc.*, **47**, 164 (1951)
73 BASOV, V.A., MARKHEVKA, V.I., MELIK-AKHNAZAROV, T.K. and OROCHKO, D.I., *Int. Chem. Eng.*, **9**, 263 (1969)
74 BEHIE, L.A. and KEHOE, P., *A.I.Ch.E. Jl*, **19**, 1070 (1973)
75 BEHIE, L.A., BERGOUGNOU, M.A. and BAKER, C.G.J., in *Fluidization Technology*, Ed. Keairns, D.L., 261, Hemisphere Publishing Co., Washington, DC, (1976)

76 DAVIDSON, J.F. and HARRISON, D., *Fluidized Particles*, Cambridge University Press (1963)
77 ROWE, P.N., MACGILLIVRAY, H.J. and CHEESMAN, D.J., *Trans. I. Chem. Engrs* **57**, 194 (1979)
78 GEORGE, S.E. and GRACE, J.R., *A.I.Ch.E. Symp. Ser.*, **74**(176), 67 (1978)
79 DE LASA, H.I., ERRAZU, A., BARREIRO, E. and SOLIOZ, S., *Can. J. Chem. Engng*, **59**, 549 (1981)

CHAPTER 4

Combustion and gasification

4.1 Introduction

Research into the combustion of fuels in fluidized beds began in earnest in the early 1960s[1] but it is only in the last few years that widespread attention has been given to the subject. This upsurge of interest has been brought about by two factors: (i) the growing shortage and increasing cost of high-grade fossil fuels, (ii) the introduction in several countries of legislation to limit further the emission of solid and gaseous pollutants from coal- and oil-fired power stations. A fluidized bed offers a means of burning coal and other solid fuels efficiently but with smaller equipment and with less pollution than more conventional methods of combustion, and so provides an attractive alternative design for power generation purposes. The advantages of burning oil or gas in a fluidized bed are not so marked however since this combustion depends strongly on the efficiency of gas mixing which is not normally high.

If coal combustion has been one of the more recent applications of fluidization technology, coal gasification in the form of the Winkler process (1922) was one of the earliest. In Europe a considerable amount of research was carried out into fluidized gasification processes and several reached the commercial scale, but with the discovery around 1960 of large deposits of natural gas, first in the Netherlands and then under the North Sea, they became obsolete. Since the energy crisis of 1973 however there has been a revival of interest particularly in the USA, and fluid bed gasification processes are once again under active investigation.

4.2 General features of fluidized coal combustion

The combustion of coal in a fluidized bed is normally carried out by continuously feeding coal parricles to a bed of some inert material such as sand or coal ash maintained at a temperature of between 750 and 900°C and fluidized with air. The steady-state concentration of coal in the bed is maintained at a low value, typically 0.5 per cent. Bed temperature is controlled by removing the heat of combustion with immersed cooling coils supplied with water, steam or air. On entering the bed the coal particles begin to devolatilize, the volatile components burning both in the bed and, under certain conditions, in the freeboard region above it. Coal particles move about freely within the bed and in general burn at a temperature somewhat above the bed average.

The features of this process that make it an attractive alternative to conventional combustion systems are as follows:

(i) *Combustion temperature*
The relatively low operating temperature is advantageous in several ways. Firstly it prevents the fusion of coal ash, leaving it soft and friable, resulting in a considerable reduction in the erosion of exposed heat transfer surfaces. (In conventional pulverized fuel (PF) combustion, erosion is caused by the sintered ash particles formed at the flame temperature of around 1500°C.) Secondly it is too low for the corrosive alkali metal salts, silica, vanadium, etc. present in the coal, to be volatilized to the same extent as in PF burning; thirdly it limits the formation of oxides of nitrogen (NO_x).

(ii) *Heat transfer*
Bed-to-immersed-surface heat transfer coefficients are very high (150–400 W/m^2K) and are up to an order of magnitude greater than those common for gas-to-surface heat exchangers. This can result in a considerable reduction in equipment size compared with conventional systems.

Careful design of coil heat exchangers for fluid bed operation is however essential in view of the high temperature gradients observed in combustors with horizontally mounted coils[2]. Too dense a packing of coils can hinder the rapid circulation of particles essential for the maintenance of a more-or-less isothermal bed although vertically mounted coils appear to be more favourable in this respect. There have also been reports of variations in bed-to-surface heat transfer coefficients with coil position, a coil located near the distributor grid measuring a lower transfer coefficient than one placed higher up in the bed[1,2].

(iii) *Gas turbine operation*
The feasibility of linking a gas turbine to a fluid bed coal combustor stems from the previously mentioned advantages of low temperature operation. Gas turbines cannot be used in conjunction with a PF burner because of the erosion and corrosion of the turbine blades that would result from using as the working fluid exhaust gases containing sintered ash and volatilized alkali metal salts. These materials are absent from the off-gases of a fluid bed coal burner and so, by operating the bed under pressure, these 'clean' exhaust gases may be expanded through a gas turbine and used directly for power generation. Moreover, operation at elevated pressures results in a higher rate of combustion since this is largely controlled by the rate at which

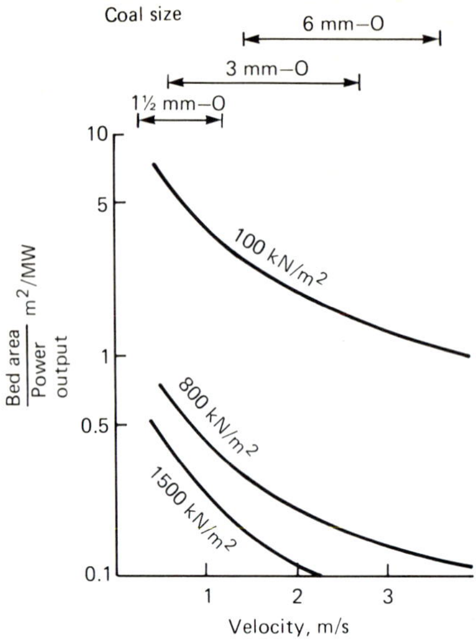

Figure 4.1 Area of bed per unit power output (From Roberts et al.[6])

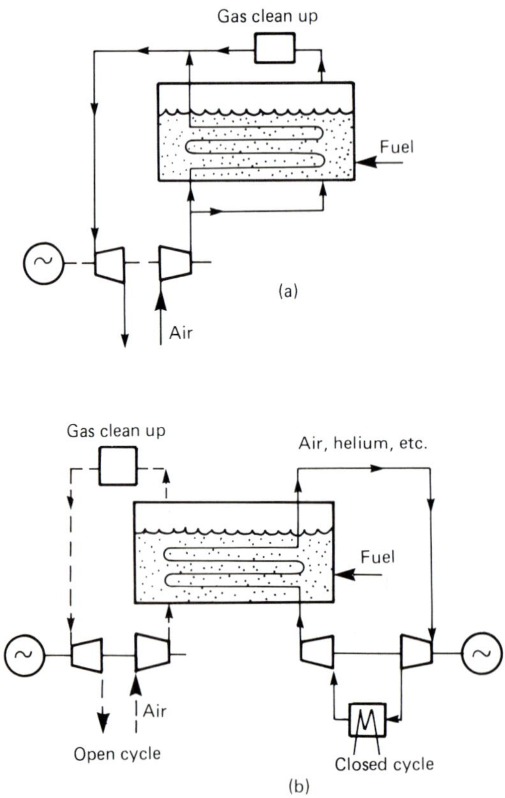

Figure 4.2 Forms of operation of fluidizied-bed combustors (From Roberts et al.[6]). (a) Open-cycle gas turbine; (b) supercharged closed-cycle gas turbine

air is supplied to the bed. A further reduction in unit size is thus possible (*Figure 4.1*) although deeper beds are required to accommodate the extra heat transfer surface. Two forms of cycle operation are illustrated in *Figure 4.2*.

(iv) *Control of SO_2 emission*
By adding particles of limestone or dolomite to the bed material, the SO_2 produced during coal combustion may be converted to calcium sulphate and retained in solid form. This has opened the way to the efficient, pollution free combustion of high-sulphur coals.

(v) *Low-value fuels*
Fuels such as anthracite fines, lignites, oil shale and coal washery tailings which are largely incombustible by conventional methods may be burned readily in a fluid bed system.

(vi) *Burning rates*
Skinner[1] has reviewed work carried out on a pilot-scale combustor in which heat release rates in excess of 2.52 MW/m^2 based on distributor area were attained. This was calculated to be equivalent to 2.07 MW/m^3 (based on total furnace volume) for a combustor burning unwashed coal with a calorific value of 23.3 MJ/kg, a burning rate which compared favourably with that attainable in a contemporary (1971) chain-grate stoker (*c.* 0.9 MW/m^3). Modern designs of chain-grate stokers however have burning rates that are comparable with but still less than those available from a fluid bed combustor[3] (2.0 MW/m^2 compared with 3.0 MW/m^2).

4.3 Plant developments

The design elements of a fluidized coal combustor were discussed by Skinner[1] on the basis of the results of small-scale experiments carried out over a period of years by organizations in the UK (National Coal Board (NCB), Central Electricity Generating Board (CEGB), British Coal Utilisation Research Association (BCURA)); the USA (Bureau of Mines, Combustion Engineering Inc., Pope, Evans and Robbins) and Australia (Commonwealth Scientific and Industrial Research Organisation (CSIRO)). These programmes were concerned with demonstrating the viability of fluidized coal combustion in competition with conventional methods and in the process all the common operating variables were investigated and their effects on combustion efficiency, heat transfer, rate of heat release, tube corrosion, etc. were studied.

The original aims of the development programmes in the UK and the USA were, as Elliott has pointed out[4], quite different, the British being concerned with potential reductions in the capital costs of power stations while the Americans placed more emphasis on pollution control. Both programmes have however progressed along similar lines. In the UK the

NCB sponsored the development of three pilot-scale units, a 1 MW water-tube boiler operating at atmospheric pressure, a 2.5 MW vertical shell boiler, and a 2 MW unit operating at elevated pressure (up to 6 bar) for studies of combined cycle power generation. Several reports describing the operation of these units have been published[3,5]. The NCB work showed the importance of balancing the requirements of a deep bed to contain the necessary amount of heat transfer surface against the penalty incurred in providing more powerful air blowers than would be required by a shallower bed. The effects of the sizes of bed and coal particles were investigated as also was the fluidizing gas velocity and its influence on combustion efficiency and particle elutriation. It was demonstrated clearly that combustion efficiency, measured by the percentage of coal burnt to carbon dioxide within the bed, was strongly dependent on maintaining a low concentration of coal in the bed and operating at as high a temperature as possible but at no more than 200–300°C below the initial deformation temperature of the ash, i.e. a maximum of about 900°C (*Figure 4.3*).

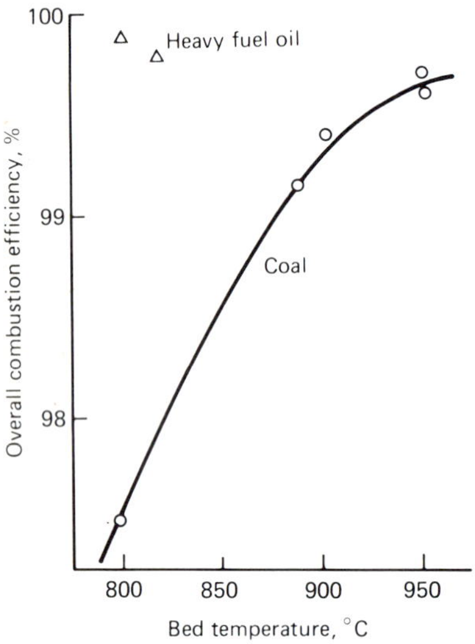

Figure 4.3 Effect of temperature on combustion efficiency at 500–600 kN/m² with 15–20% excess air (From Roberts et al.[6])

The problem of combustible gases such as CO being emitted into the freeboard is serious, since it can lead to uncontrollable temperature increases above the bed which can nullify the advantages of low-temperature operation. Roberts *et al.*[6] found evidence of freeboard combustion when burning coal at bed temperatures of 800°C and below but none at higher temperatures and none when burning oil.

Several studies have indicated that the most important single factor influencing combustion efficiency in a fluid bed is the carbon lost in the form of elutriated particles. The results of the BCURA experiments[1] on solids carry-over from a 686 mm diameter bed showed that carbon elutriation increases with coal feed rate and fluidizing gas velocity but that the loss may be reduced by increasing the excess air and the recycle rate.

Similar conclusions have been arrived at more recently by Gibbs *et al.*[7] from a comprehensive programme of experiments with a 300 mm square bed. The results of their measurements of carbon loss as a function of excess air over a temperature range of 770–840°C indicated that with excess air fractions of over about 25 per cent, the major loss was accounted for by elutriated particles which was estimated to be equivalent to some six per cent of the potential heat available in the coal. Vogel *et al.*[8] found from experiments conducted at atmospheric pressure that 10 to 20 per cent of the unburned carbon was in the form of CO and methane in the flue gas but that when the experiment was repeated at 8 bar pressure this loss was reduced to 0.2 to 1.3 per cent. Carbon elutriated in the fly ash may be either recycled[1] or burnt in a second bed operating in series[9]; the latter method has been shown to result in combustion efficiences of over 99 per cent[10].

The NCB work also demonstrated the importance of coal distribution and showed that high combustion efficiency depends on achieving a rapid and uniform distribution of coal over the cross-section of the bed. Although vertical mixing was generally very good, lateral mixing could present a considerable problem, which was only solved by employing a large number of coal-inlet nozzles. Another problem to which no completely satisfactory solution yet appears to have been found is that of operating a fluid bed combustor over a wide load range. The problem arises from the fact that the major factor controlling heat generation in the bed is the bed temperature and this is variable over only a narrow range.

The experience of the American company, Pope, Evans and Robbins (PER) in developing fluidized coal combustors has been documented in a number of publications; Bishop[11] has described the programme that led to the development of the modular fluidized bed boiler (FBM) incorporating a carbon burn-up cell (CBC) for dealing with fuel particles elutriated from the main bed. Further details have been provided by Ehrlich[12] who also discussed the problem of load variation and concluded that this may be achieved only by satisfactorily varying the mass of the active bed. Turn down would thus be carried out by reducing the number of modules operating in the fluidized mode. The same thinking lies behind the Westinghouse design for a 600 MW(e) unit[13] which is based on a system of four vertically mounted pressure vessels each containing four separate fluid beds and operating as a mixed steam/gas-turbine cycle plant.

In 1975 agreement was reached between the governments of the UK, USA and the Federal Republic of Germany to share the cost of a large pressurised fluidized bed combustion facility at Grimethorpe in the UK. The unit was completed in 1980 and at the time of writing is undergoing a test programme of some 2900 hours having satisfactorily completed a 300 hour commisioning run. The fluidized bed is contained within a 2 m square chamber through the walls of which cooling water is circulated; the chamber itself is housed inside a pressure vessel 14 m high and 4 m in diameter (*Figure 4.4*). Crushed coal enters the bed through nine feed points situated in the base and combustion is carried out at pressures of up to 12 bar; steam is generated via water-fed tubes immersed in the bed. Coal combustion rates of up to 10 tonne/hour are achieved giving power outputs of up to 65 MW. The off-gases from the bed are cleaned by four parallel trains of two cyclones and a major objective of the current test programme is the assessment of the effects of these gases on turbine blades. The incorporation of an industrial gas turbine into the facility, which is operated under the auspices of the International Energy Agency, is planned for a later stage.

Before concluding this Section, mention must be made of the Ignifluid boiler developed by the French engineer A.A. Godel[14]. It represents the first large-scale application of fluidized combustion technology and has been operated successfully at a number of French power stations and at Casablanca, Morocco for a number of years[15]. In design the Ignifluid boiler is midway between the chain-grate stoker and the normal fluidized bed combustor. It consists of a travelling, inclined grate which supports a fuel bed fluidized by a current of primary air. Combustion is completed by a stream of secondary air blown over the bed surface; since the fuel burns throughout the full depth of the bed (some 60 cm) high rates of combustion are achieved. Combustion temperatures are higher than those being considered for fluidized bed boilers (1200°C compared with 850°C) and as a result the coal ash fuses to form clinker which is removed by the action of the chain grate. A wide range of fuel types from anthracite to coking bituminous coal may be used and the combustion efficiency is comparable with that of PF burning. Indeed the Ignifluid boiler would seem to offer many of the advantages over conventional boilers that are claimed for fluidized bed combustors and, with more than a decade of proven large-scale operation behind it, the system must constitute a serious competitive threat to the first generation of low-temperature fluid bed units. Ignifluid boilers were, however, developed to burn low-sulphur European coals and their ability to treat fuels with high sulphur contents without producing a correspondingly high emission of sulphur dioxide has yet to be demonstrated.

4.4 The mechanism of coal combustion

The combustion of coal is a complex process and involves the interplay of chemical reactions with heat and mass transfer effects. In a fluidized bed the added complications of bed hydrodynamics are also present and it is fair to say that at the present time no complete understanding of the process has been achieved.

Much attention has however been given to the combustion of carbon in air and oxygen and some of this work is relevant to the present problem. The early study by Tu, Davis and Hottel[16] established experimentally the

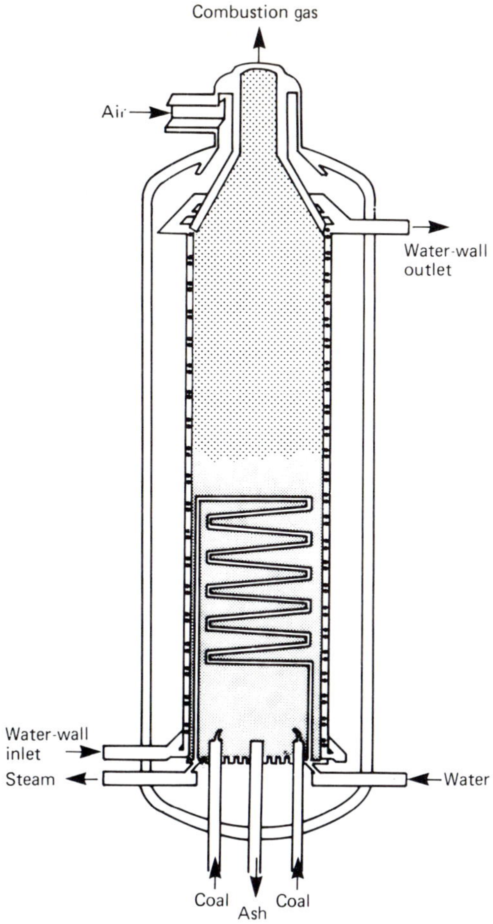

Figure 4.4 The Grimethorpe pressurized fluid bed combustor

linear relationship between burning time and the square of the initial diameter of a carbon sphere. This had been suggested on theoretical grounds by Nusselt[17], whose proposed combustion mechanism involved a so-called 'single-film' model in which oxidation occurs by diffusion of oxygen through a stationary film surrounding the carbon surface where it reacts to form CO and CO_2 (*Figure 4.5(a)*). An alternative 'double-film' model, proposed initially by Burke and Schumann[18] assumes the primary product of the oxidation to be CO which diffuses away and burns to CO_2 in a homogeneous gas phase reaction in a zone close to the carbon surface. Half the CO_2 thus formed diffuses away from the particle while the rest flows back to the surface and reacts to form CO. The steady-state concentration profiles on the basis of the double-film theory are as shown in *Figure 4.5(b)*. The two models have received support from a number of investigators and useful reviews of the subject are to be found in the works of Spalding[19], Field *et al.*[20], Mulcahy and Smith[21], Smith[22] and Caram and Amundson[23].

Following the pioneering work of Avedesian and Davidson[24] on the combustion of char particles in a laboratory-scale fluidized-bed combustor, a number of similar studies have been carried out with the objective of establishing the detailed mechanism of coal combustion in fluid-beds of small particles. It has been pointed out however that such studies may have little relevance to the processes taking place in industrial-scale units where relatively large particles are fluidized at high gas velocities. As Park *et al.*[25] have observed, the hydrodynamics of such large-scale systems are more likely to involve cloudless bubbles, moving slowly relative to the interstitial gas flow, than the clouded fast-moving bubbles typical of laboratory-scale equipment, and therefore the combustion mechanisms in the two cases are likely to be different. While this is almost certainly true in the case of power-generating units operated at, or near, atmospheric pressure, there are bound to be industrial situations in which the more familiar fast-moving bubbles, or some variant of them, are the dominant features, and so interest in the study of small-scale fluidized-bed combustors is not wholly academic. We will therefore examine the work carried out on the laboratory scale with beds of small particles before turning to the question of combustion in large particle beds.

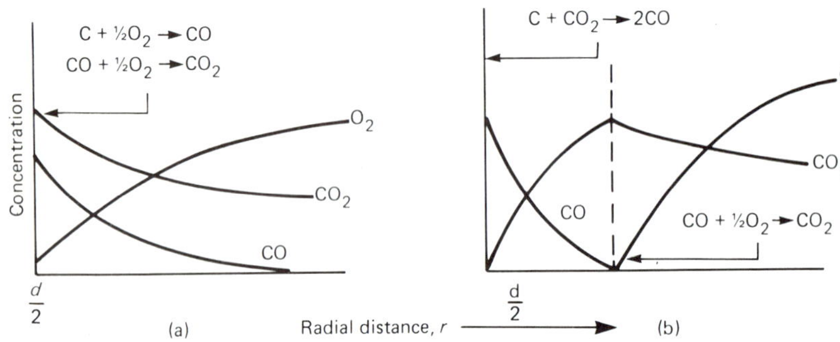

Figure 4.5 Mechanisms of char combustion in a fluidized bed; (a) single film kinetic/diffusion control, (b) double film diffusion control

Figure 4.6 summarizes the sequence of events that is observed when a piece of coal, say 8 mm across, is put into a fluidized bed of small sand particles, some 400 μm in diameter, at a temperature in the region of 800°C:

(1) Unless steps are taken to force the coal to mix with the bed solids (e.g. by 'Gulf Streaming'–Chapter 1, Section 1.10.1) it will remain on or near the bed surface for a short period of time during which it loses its volatile components. These volatile materials burn as a diffusion flame surrounding the particle and appear to be responsible for maintaining it at the surface. During this period the coal appears black against the red radiation field of the bed, indicating that its temperature, T_p, is somewhat lower than that of the bed itself, T_B.

(2) When devolatilization is complete the resulting particle of char sinks into the bed where it burns in contact with the fluidizing air at a temperature rather higher than T_B. During this period, which is of much longer duration than (1), the particle appears to move freely within the bed only rising to the surface intermittantly, when its colour temperature is seen to be in excess of T_B.

(3) When the burning char has shrunk to a critical size it is carried by the fluidizing air out of the bed and into the freeboard where its temperature rises rapidly and it either burns away to an ash fragment or is transported from the bed and quenched.

4.4.1 Devolatilization and volatiles combustion

The volatile component of coal can be an important factor in the overall combustion process, accounting as it does for a high proportion (sometimes as much as half) of the heating value of the fuel[26]. It is therefore surprising that relatively little attention has been paid to the mechanism of release of volatiles from coal and to their subsequent reaction in the fluidized bed.

Borghi *et al.*[27] made theoretical predictions of the rate of release of volatiles from a bituminous coal and compared these with the burning rate of the resultant char. They assumed the rate of combustion of the volatiles was infinitely fast and found the devolatilization time to be two orders of magnitude lower than the burn-out time of the corresponding char, i.e. 10 s compared with 1000 s for a 1.5 mm coal particle at 1023 K.

An experimental programme designed to measure the burn-out times of volatiles, t_v, as a function of coal particle size, d_i, volatile content and mass of charge, m, was described by Atimtay[28]. She showed that for a particular coal t_v was a linear function of both d_i and m, the volatiles from a coal with 33.8 per cent volatile matter (dry basis) and particles ranging from 1.52 to 3.06 mm taking 3–12 s to burn. Furthermore, on the basis of a model that assumed the volatiles to be emitted in the form of a large number of radial jets, and to burn on contact with inward diffusing oxygen in a reaction zone around the particle, Atimtay obtained an expression for the radial velocity of the evolved material, V_R:

$$V_R = \frac{Q}{\pi d_i^2 N} \tag{4.1}$$

where $Q = V/t_v$ is the volumetric flowrate of the volatiles at the bed temperature and N is the number of particles in a batch. Radial velocities calculated on the basis of equation 4.1 were found to be of the same order of magnitude as the minimum fluidization velocity of the coal particles and it was suggested that this could account for the observation that the particles float on the bed surface during the devolatilization period.

The mechanism of volatiles release from a stationary source was

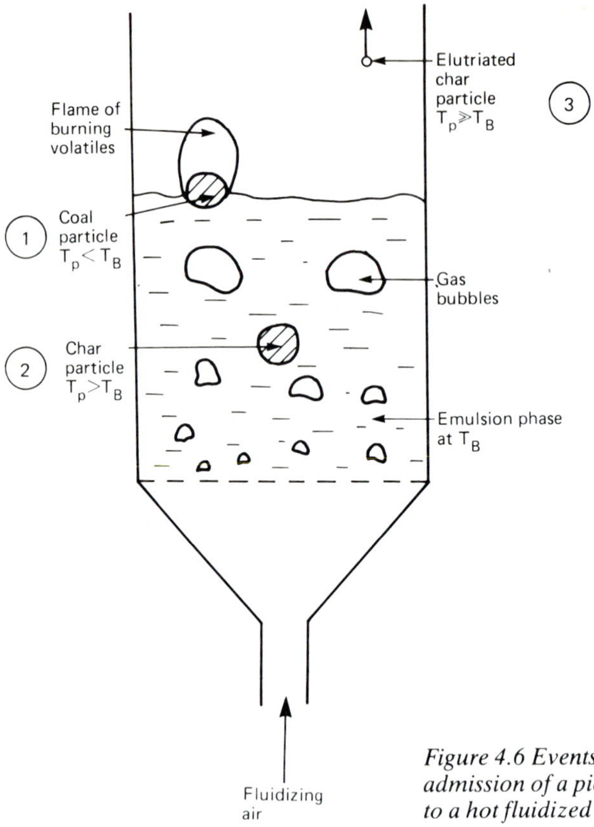

Figure 4.6 Events following the admission of a piece of bituminous coal to a hot fluidized bed of sand

examined by Yates *et al.*[29] who suspended a piece of bituminous coal in a fluidized bed at 850°C and observed by means of X-rays the formation and development of a stream of bubbles.

Figure 4.7 shows a sequence (1–6) from a cine film of the event taken at a speed of 50 frames/s. The film was taken some 10 s after the coal was put in place, i.e. at the end of the period required for the coal to reach bed temperature and before the loss of its volatile components was complete. The vertical supporting wire can be seen on the right of the field of view, while the electrical heating wires appear as dark horizontal lines; the coal, which was transparent to X-rays, was held in the horizontal loop at the bottom end of the supporting wire.

Frame 1 shows a large air bubble leaving the bed at the top left hand corner, and a wisp of gas beginning to emerge from the coal. In Frames 2 and 3 the void rising from the coal can be seen more clearly while in Frame 4 it has begun to take on the familiar form of a gas bubble. Frames 5 and 6 show the bubble from the coal growing and moving ahead of the next air bubble which is rising from the distributor. Visual observation of the coal (by means of a scintillation screen) showed a stream of such bubbles rising continuously for about one minute after its immersion in the bed. At the end of this time no more bubbles could be seen and it was assumed that the coal had by then become devolatized and had begun to burn as char.

The bubbles appeared as discrete volumes of burning gas at the bed surface and their evolution ceased after a period of about one minute. The combustion of the volatiles is this case would appear to have followed the same pattern as that of any other combustible gas admitted as bubbles to the lower regions of a hot fluidized bed.

The problem of burning industrial waste materials that have a high content of volatiles has been discussed by Ormston *et al.*[30], who used a variant of the 'Gulf Streaming' technique (Chapter 1, Section 1.10.1) to ensure the burning particles were well mixed with bed solids and did not accumulate at the surface. In their design, circulation of bed particles is induced by providing an unequal distribution of fluidizing air across the distributor (*Figure 4.8*). In practice the air supply is split into four components, three of which flow into the divided plenum chamber and one over the bed surface. The combustible waste, consisting of acid tars and waste oils, is introduced through a pipe situated in the bed wall at the point of maximum downward motion of the bed solids, and the volatile

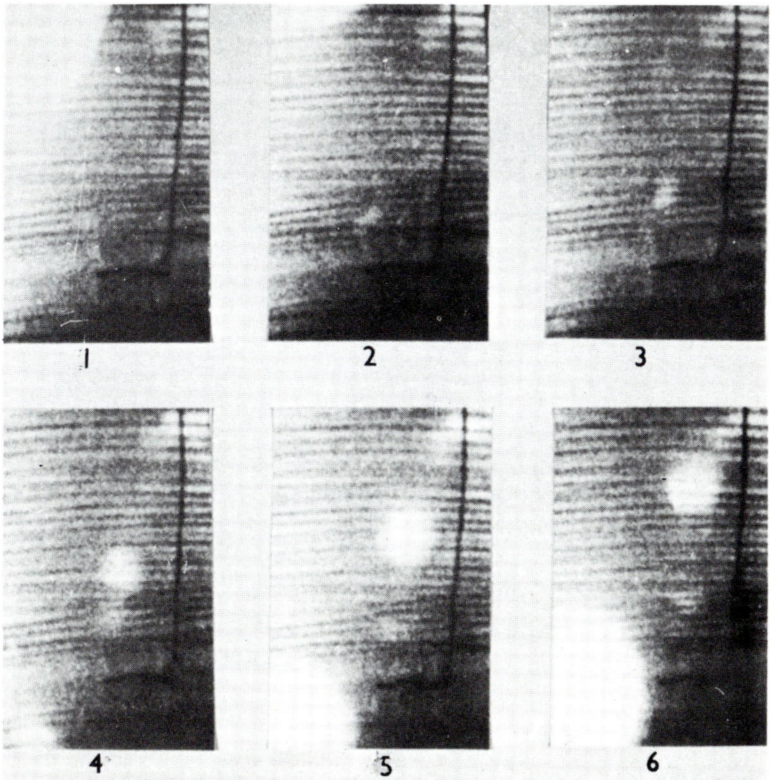

Figure 4.7 X-ray photographs of a devolatilizing coal particle in a fluidized bed of sand ($d_p \simeq \mu m$) at 850°C (From Yates et al.[29])

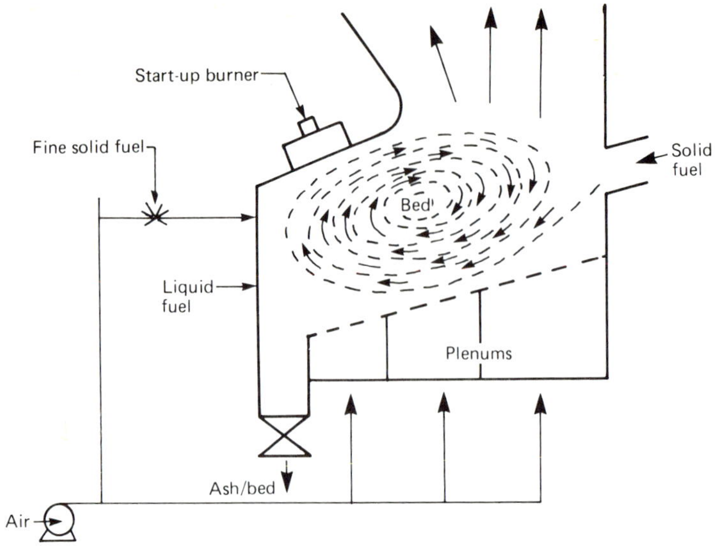

Figure 4.8 Circulating fluidized bed (From Ormston et al.[30])

components are burned in a controlled manner within the bed. A plant of this type has operated for some years at Avonmouth, England, where it is used not only to dispose of difficult wastes but at the same time to raise 3200 kg/hr of steam.

The combustion of volatiles in beds of large particles, such as might be used in electrical power generating systems, has been investigated by Park *et al.*[25]; this work is considered in detail in Section 4.4.4.

4.4.2 Char combustion

The study by Avedesian and Davidson[24] was one of the first attempts to model the combustion of char particles in a fluidized bed. Their theoretical treatment was combined with an experimental study of the combustion of batches of char particles of diameter 0.23 to 2.61 mm in a bed of coal ash in which burn-out times and off-gas oxygen and carbon dioxide concentrations were measured under a variety of conditions. The fundamental assumption made by Avedesian and Davidson, based on previous work by other authors[19,20,31], was that the combustion rate is controlled by the rate of diffusion of oxygen to the burning particle and that the chemical reactions are so fast at the temperatures in question that they have no influence on the overall kinetics. They therefore proposed that the two-film mechanism operated in a fluidized bed combustor with carbon monoxide first being formed at the char particle surface by the reduction of carbon dioxide in the Boudouard reaction, $C + CO_2 = 2CO$, and then diffusing away to burn on contact with oxygen in a flame surrounding the particle. Although this picture is now thought to be incorrect and that the chemical reactions are not as fast as was supposed, it is useful to follow the reasoning of Avedesian and Davidson through to its conclusion and then to see how it can be modified in the light of subsequent work to give a more realistic description.

The equation of continuity for a species of concentration C in a spherically symmetrical system with radial distance r is:

$$\frac{d}{dr}\left[r^2\left(\frac{dC}{dr}\right)\right] = 0 \tag{4.2}$$

This can be solved with appropriate boundary conditions to give expressions for the concentration profiles of the diffusing species in the two zones. Since the rate of combustion is assumed to be diffusion controlled, the burn-out time of a char particle can be calculated once the flowrate of oxygen to the particle is known. The molar flowrate of oxygen was shown by Avedesian and Davidson to be:

$$n = 4\pi d_p E C_p \tag{4.3}$$

where E is an effective diffusion coefficient for mass transfer through the dense phase, d_p is the particle diameter and C_p the dense-phase oxygen concentration. A Sherwood number was defined:

$$Sh = \frac{2E}{D_G} \tag{4.4}$$

and assumed to be approximately constant although its value was expected to increase from $Sh = 2\varepsilon_{mf}$ (where ε_{mf} is the dense-phase voidage) for large char particles to $Sh = 2$ for particles smaller than the bed material. From the above equations the molar flowrate of oxygen to a particle and the carbon burning rate, $\dot{C}$ (mol/s), are:

$$n = \dot{C} = 2\pi Sh D_G d_p C_p \tag{4.5}$$

Then using an equation for the overall consumption of oxygen based on the fluid-bed reactor model of Davidson and Harrison (assuming a completely mixed emulsion phase–*see* Section 2.3.2), the authors obtained the following expression for the burn-out time of a batch in a bed of cross-sectional area A:

$$t_c = \frac{m}{12C_oA[(U - U_{mf})(1 - \exp - X) + U_{mf}]} + \frac{d_i^2\rho_c}{96 Sh D_G C_o} \tag{4.6}$$

where C_o is the inlet oxygen concentration, ρ_c the char density, d_i the initial particle diameter, m the total mass of char and X the interphase mass-transfer coefficient (*see* Section 2.2.2.1). The significance of the two terms on the right-hand side of equation 4.6 is: (a) the first gives the value of t_c

for a large charge. If m is large, C_p is very low and the burning rate is controlled by X; (b) the second term gives t_c for a single particle and shows that for diffusion control of reaction, $t_c \propto d_i^2$.

Avedesian and Davidson studied the effect on burn-out time of (i) carbon charge, (ii) initial diameter of char particles, and (iii) char density, and found their results to be in good agreement with the predictions of the model. Campbell and Davidson[32] extended the foregoing model to describe the situation in which char is fed continuously to the bed and also reported good agreement between theory and experiment.

Now the combustion of coal particles can be treated in general on the basis of the 'shrinking core' model[33], from which it can be shown (*see* Levenspiel[34]) that under conditions of chemical control of the burning rate, the burn-out time for a single particle undergoing the reaction $bC + O_2 \rightarrow$ products, is :

$$t_c = \frac{\rho_B d_i}{2bk_c C_A} \tag{4.7}$$

where ρ_B is the molar density of the solid, k_c is a surface rate constant and C_A is the gas phase oxygen concentration. In other words, for the region of chemical control, $t_c \propto d_i$.

Tomeczek pointed out[35] that the results of Avedesian and Davidson gave straight lines when t_c was plotted against *both* d_i and d_i^2, and he concluded that the experimental method was not sufficiently sensitive to distinguish between diffusion and chemical control of the reaction. Further doubt was cast on the Avedesian–Davidson interpretation by Borghi *et al.*[27] and by Basu *et al.*[36], both groups questioning the assumption of a fast rate for the Boudouard reaction at the temperatures of a fluid-bed combustor. Moreover Basu *et al.* showed by means of a simple heat balance calculation that insufficient heat was available to support the Boudouard reaction which is known to be endothermic to the extent of 170 kJ/mol, and demonstrated experimentally that both CO and CO_2 were formed at the carbon surface. They therefore suggested that the single-film mechanism was more likely than the double-film mechanism of Avedesian and Davidson and proposed a modified expression for the burning rate of a single carbon particle (mol/s):

$$\dot{C} = 4\pi D_G \varepsilon_m d_p C_p \left(1 + \frac{12C_p}{\rho_g}\right) \tag{4.8}$$

where ε_m is the mean void fraction around a burning particle and ρ_g is the density of the fluidizing gas.

Chakraborty and Howard[37] measured the burning rates of particles of electrode carbon in a shallow fluidized bed and found the rate to vary according to $\dot{C} \propto d_i^n$, where $1.8 < n < 1.97$. On the basis that n should be respectively 1 and 2 for diffusion and kinetic control they concluded that the combustion was mainly kinetically controlled.

A modification to the Avedesian–Davidson model was proposed subsequently by Ross and Davidson[38] (and reported by La Nauze[39]) whose rate equation takes account of both diffusional and kinetic resistances. The rate of carbon consumption is:

$$\dot{C} = 2\pi d_p^2 K C_p \tag{4.9}$$

where the overall rate constant K is a combination of kinetic and diffusional terms such that:

$$\frac{1}{K} = \frac{\gamma}{k_c} + \frac{\alpha d_p}{Sh D_G} \tag{4.10}$$

Here k_c is a reaction rate coefficient defined as:

carbon reaction rate per unit surface area = k_c (surface concentration of oxygen).

α and γ are constants whose values depend on the reactions assumed to be occuring at the surface:

	α	γ
(i) $C + \frac{1}{2}O_2 \rightarrow CO$	1	2
(ii) $C + O_2 \rightarrow CO_2$	2	2
(iii) $C + CO_2 \rightarrow 2CO$	1	1

The resulting expression for the burn-out time of a batch is:

$$t_c = \frac{m}{12C_0A[U - (U - U_{mf})\exp - X]} + \frac{\rho_c d_i^2 \alpha}{96 Sh D_G C_0} + \frac{\rho_c d_i \gamma}{48 k_c C_0} \tag{4.11}$$

where the last term refers to conditions of chemical control (compare equation 4.6).

Ross and Davidson reported batch experiments in which the relative importance of kinetic and diffusional effects were determined[40]. Their results are given in *Figure 4.9* and compared with theoretical calculations of Borghi *et al.*[27]; from these they conclude that the combustion rate is controlled by a combination of the two effects. For large particles (d_p>3 mm) the rate is controlled by diffusion of oxygen to the surface; in the case of small particles however, combustion is controlled mainly by the kinetics of the reaction $C + \frac{1}{2}O_2 = CO$, the CO so formed then burning in the emulsion phase around the particles. In the latter case the heat of combustion of the CO is dissipated in the emulsion solids with the result

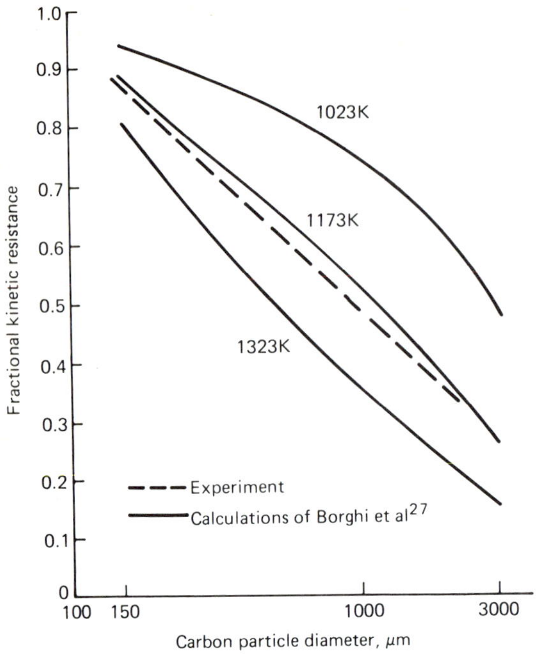

Figure 4.9 Effect of particle size on kinetic dependence of char combustion (From Ross and Davidson[40])

that small carbon particles are likely to be at or near the bed temperature. Large particles on the other hand are generally hotter than the bed.

4.4.3 Particle temperatures

The temperature reached by a burning particle is an important factor in any mechanism of the fluid-bed combustion process, particularly since it has a strong influence on the rates of chemical reactions taking place at the carbon surface. If particle temperatures are low, rates of, for example, the Boudouard reaction will also be low and the combustion could be chemically rather than diffusion controlled. On the other hand, high particle temperatures imply high reaction rates and the likelihood of diffusion control. There has been however a general lack of agreement about the actual temperatures reached by burning char particles partly as a result of the experimental difficulty of the *in situ* measurement of the temperature of a small, freely-moving object in a fluid-bed combustor.

A number of studies have been reported in which thermocouples were embedded in particles and direct temperature readings obtained. Bhattacharyya and Pei[41] measured fluid-to-particle heat transfer in this way in a low temperature bed, while Chakraborty and Howard[42,43] used the same technique in a fluid-bed combustor for particle temperature measurement. The latter workers found temperature differences between bed and burning particle ranging from 50 to 215°C, the largest difference being found for the smallest particles. The thermocouple method suffers however from the disadvantage that the freedom of movement of the particles is inevitably impaired, and as a result temperatures measured in this way are not thought to be fully representative of those reached by an unrestrained particle.

A different experimental approach was adopted by Yates and Walker[44], who embedded rings of fine heat fusible wires of accurately known melting point inside spherical particles of compressed anthracite powder and attempted to recover the rings at the end of a combustion run. If the rings were not recovered on sieving the bed material then it was assumed they had fused and that the particle had reached a temperature corresponding at least to that of the melting point of the metal. Results for three different sizes of ring melting at 1000°C over a range of bed temperatures are given in *Table 4.1*, from which the following conclusions were drawn:

> For a bed temperature, T_b, of 900–940°C the temperature, T_p, reached by the burning char particles of 7.5 mm diameter is no more than 1000°C, while for 2.5 mm diameter particles in a bed at 860–900°C, $T_p - T_b$ is in the range 100–140°C. Furthermore the particles become hotter as their diameters decrease. The temperatures indicated by this method are generally higher than those calculated by Zabrodsky and Parnas[45] for carbon particles of 1.44 mm diameter burning in a fluidized bed with 10 per cent excess air, but lower than those measured by Whellock[46] for coal particles using the thermocouple technique.

Two subsequent papers reported the use of photographic techniques to measure particle temperatures. Roscoe *et al.*[47] photographed the surface of a fluid bed at 930°C and compared the brightness of the observed burning

Table 4.1 Results of ring fusion experiments (from Yates and Walker[44])

Bed temperature (K)	*Ring diameter* (mm)	*Recovery of intact rings* (%)
	2.5	100
1030	4.5	100
	7.5	100
	2.5	100
1130	4.5	100
	7.5	100
	2.5	0
1173	4.5	100
	7.5	100
	2.5	0
1213	4.5	0
	7.5	67
	2.5	0
1233	4.5	0
	7.5	0

particles with that of a black body radiator exposed on the same negative. The films were black and white Kodak TriX. They concluded that the particle temperatures were 130–160°C above that of the bed. Ross *et al.*[48] used both black and white and colour photography of a bed surface to measure particle temperatures as a function of the oxygen concentration of the fluidizing gas. They found the relationship:

$$T_p - T_b = 0.66\, C_p \tag{4.12}$$

Here the dense phase oxygen concentration C_p (mol/m^3) was determined from the Davidson–Harrison fluid-bed reactor model (*see* Section 2.3.2) with a completely mixed emulsion phase:

$$\dot{C} = A(C_o - C_p)[U - (U - U_{mf})\exp - X] \tag{4.13}$$

The rate of combustion, $\dot{C}$, was calculated from off-gas CO_2 analysis and the interphase transfer factor, X, was found empirically to have a value of 2.2. The predictions of equation 4.12 agree well with the measured values of both Chakraborty and Howard, and Yates and Walker, although the authors warn that the relationship may not apply at bed temperatures different from 900°C and may not be universal for all carbons.

4.4.4 Bubble models and the 'plume' model

The model of Avedesian and Davidson and its subsequent modifications discussed in Section 4.4.2 are among a number of mechanistic models of fluidized bed coal combustion that have been developed in recent years and, as with the models of chemical reactors discussed in Chapter 2 they differ according to the combination of assumptions they make. Some for instance are based on the simple two-phase theory of fluidization while others take account of bubble clouds and wakes; some attempt to be fully comprehensive and allow for volatiles combustion separately from char combustion, while others combine volatiles and char and treat them in terms of an overall burning scheme; some models allow for carbon elutriation, others ignore it. A comparison of the major assumptions of the models has been made by Park *et al.*[25] and a modified version of their summary is given in *Table 4.2*.

In making this comparison, Park *et al.* drew attention to a serious shortcoming of the models when applied to the atmospheric pressure combustors that are currently under evaluation for industrial-scale power generation. Their reasoning was that owing to the very high throughput of fluidizing air required for a compact design, the bed particles would have to be much larger than those in normal use, perhaps up to 5 mm diameter, and that under these conditions the bed would be in either the turbulent-churning or slow-bubble flow regimes. A rough calculation illustrates the magnitude of the problem. A 250 MW(e) power generator would need to

Table 4.2 Models of fluidized-bed combustion (from Park *et al.*[25])

Reference	*Authors*	*A*	*B*	*C*	*D*	*E*	*F*
24	Avedesian and Davidson	1	1	1	3	1	No
32	Campbell and Davidson	1	1	1	3	1	No
49	Gibbs	1	1	1	1	1	Yes
50	Gordon and Amundson	1	1	1	3	⅓	No
51	Gordon *et al.*	1	1	1	3	3	No
52	Chen and Saxena	2	3	1	3	1	Yes
53	Horio *et al.*	1	1	1	3	2	Yes
54	Horio and Wen	3	2	2	1	2	Yes
55	Rajan *et al.*	3	2	2	1	2	Yes
56	Baron *et al.*	¼	¼	1	2	2	Yes
57	Baron *et al.*	1	1	1	3	1	Yes
58	Fan *et al.*	1	1	3	3	1	No
25	Park *et al.*	4	4	1	4	2	Yes
39	Ross and Davidson	1	1	1	3	⅔	No
59	Donsi *et al.*	1	⅓	1	½	2	Yes

Key

A Model for fluidized bed
1 Two phase bubbling bed model: bubble phase and emulsion phase.
2 Three phase bubbling bed model: bubble, cloud and emulsion phases.
3 Compartments in series model: two phases in each compartment.
4 Slow bubble regime model: no distinction is made between the bubble phase and the emulsion phase.

B Gas flow pattern in the bed
1 Plug flow in the bubble phase, mixed flow in the emulsion phase.
2 Mixed flow in both phases.
3 Plug flow in both phases. Gas exchange between two phases is finite.
4 Plug flow through the bed. No distinction between phases.

C Movement of solids in the bed
1 Well mixed.
2 Well mixed in each compartment.
3 Dispersion of solid is finite. Dispersion coefficient is used for mass balance.

D Devolatilization of feed coal
1 Instantaneous devolatilization. Volatiles evolve uniformly across the plane representing coal feed point.
2 Devolatilization is slower than mixing of the feed coal. Volatiles evolve uniformly throughout the bed.
3 Not considered.
4 Devolatilization is instantaneous at the feed point.

E Kinetics of solid coal combustion
1 Film diffusion controlling.
2 Both film diffusion and surface reaction influence the rate, but homogeneous oxidation of CO to CO_2 is assumed to be very fast.
3 Rate of CO oxidation is comparable to that of other reactions.

F Is elutriation considered?

burn some 800 MW of coal equivalent. Assuming the coal to have a calorific value of 3×10^4 kJ/kg the rate of coal combustion would be about 27 kg/s and would require some 300 kg/s of air. The volumetric air flowrate would thus need to be of the order of 1000 m^3/s at 900°C which, with a superficial fluidizing velocity of 3 m/s would require a total bed area of some 330 m^2, possibly divided among, say, three beds of 26 m × 4 m each. Clearly the larger the U_{mf} value of the bed material the lower the bed area required but the more likely that the particles will be of the Geldart Group D variety (Section 1.3), which fluidize in the 'slow bubble' regime.

Therefore Park *et al.* have proposed a new model, the 'plume' model, for these large-particle systems. The model is based on the assumption that since coal devolatilization is fast (1 s for a 200 μm particle[60]) compared with both particle circulation (*c.* 15 s for complete mixing[25]) and char combustion (60 s for a 300 μm particle[25]) the coal particles entering at the distributor will devolatilize practically instantaneously and the volatiles will rise in plug flow as plumes with the same velocity as the fluidizing air. Thus a plume of volatiles is envisaged as rising from each coal entry port and burning with air at the plume boundary as in the case of a diffusion flame (*Figure 4.10*). This combustion, accounting for some 50 per cent of the

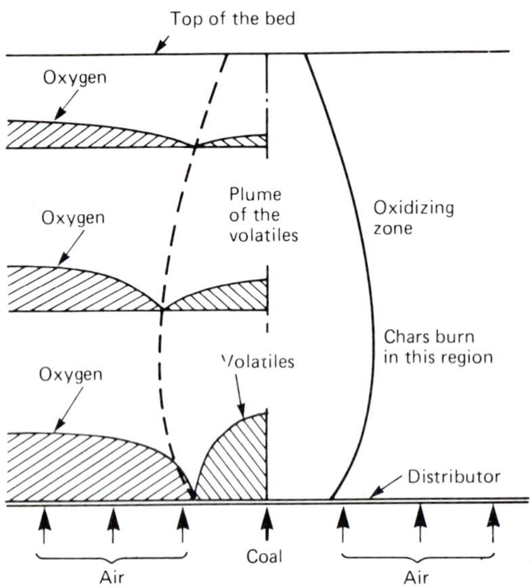

Figure 4.10 Plume in the bed (From Park et al.[25])

heating value of the coal, can be completed either in the bed or, if the plume breaks through the bed surface, in the freeboard, the latter circumstance causing a temperature jump in the gas space above the bed. Either way, volatiles combustion will lead to a depletion of oxygen in the bed and will thus have a direct bearing on the rate of combustion of the remaining char. The conversion of volatiles in the bed is governed by the dimensionless 'plume group', $HD_r/(UL^2)$, for sufficiently large values of which the volatiles burn completely in the bed (*Figure 4.11*). In this group H is the bed height, D_r a radial gas dispersion coefficient and L the radius of a coal feed module. Little experimental data have yet been published with which to test the plume model and yet it would seem to represent a step forward in our understanding of the mechanics of fluidized bed coal combustion, particularly in systems of large particles.

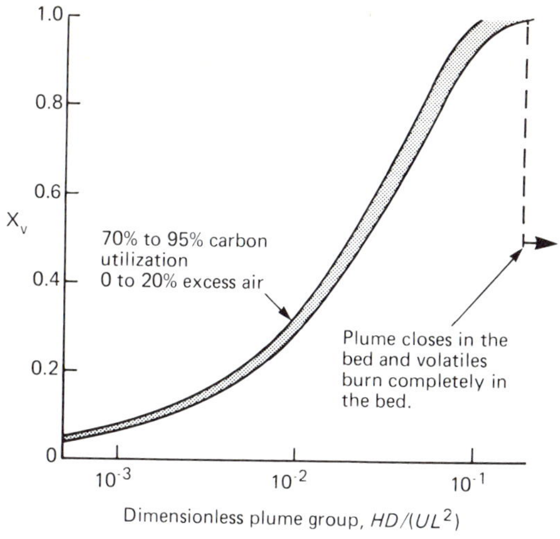

Figure 4.11 Fraction of volatiles (X_v) burned in the bed (From Park et al.[25])

4.4.5 Carbon elutriation

Although under some conditions of operation carbon monoxide and coal volatiles can be lost from the bed, the major cause of combustion inefficiency in fluidized bed systems is the loss of small unburnt particles of char or 'carbon fines'. The loss can be as much as 25 per cent of the carbon input[39]. It has been realized for some time that the extent of this loss is a function of both the carbon loading in the bed[61] and of the operating conditions[1,62–64]. An indication of this latter feature is given in *Figure 4.12* where the influence of excess air is shown. Donsi *et al.*[65] investigated these effects using a fluidized bed of 160 mm diameter and showed that carbon loss was due to the elutriation of solid particles that originated in two ways:

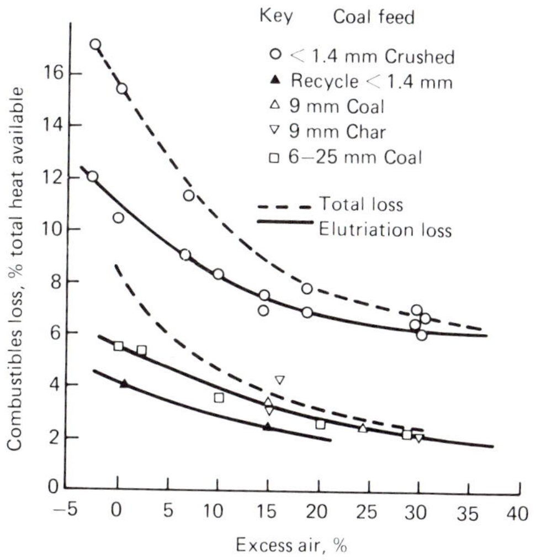

Figure 4.12 Influence of excess air on combustibles loss (From Gibbs and Hedley[64])

(a) from the unburnt cores of the coal feed, and (b) from the attrition of coal during its combustion. The contribution of (b) was dominant for the larger coal particles investigated, and a simple linear relationship was found to relate the amount of elutriated carbon from this source, E_a (g/min), with the carbon loading of the bed M_c (g):

$$E_a = E_c - E'_c = \frac{k(U - U_{mf})M_c}{\bar{d}} \quad (4.14)$$

where E_c = total rate of carbon elutriation (g/min)
E'_c = rate of carbon elutriation from unburnt cores (g/min)
$\bar{d}$ = surface mean diameter of the burning coal particles (cm)
k = an empirical constant

Some of the causes of coal particle attrition are:
(i) thermal shock resulting from rapid heating of the feed
(ii) abrasion by particle/particle and particle/surface collisions
(iii) breakage due to rapid evolution of volatiles.
Despite recent work in this area[66] much still remains to be learned about the mechanisms of these processes.

In general it would seem that once fine particles have been produced in a fluidized bed combustor their elutriation may be predicted on the basis of the empirical correlations determined for 'cold' systems (*see* Gibbs[49], Beer *et al.*[66], and Section 1.9).

4.5 Oil and gas combustion

The combustion characteristics of liquids and gases in a fluidized bed are quite different from those of solids. Provided the bed temperature is in excess of about 700°C, devolatilized coal char or coke will burn at all points within the bed and at a rate determined by the rates of reaction and diffusion of oxygen to the solid surface. For a liquid or gas to burn however, it is essential for it to be mixed intimately with oxygen or air and for the mixture to be within the flammability limits of the particular fuel. In general of course liquids themselves do not burn but must first be vapourized and the vapour mixed with oxygen in the correct proportions. There are thus many similarities between the combustion characteristics of a fuel oil and a gas and it will be convenient to consider both as gases from the point of view of fluidized bed combustion.

The mixing of fuel and air may be carried out either in the plenum chamber or windbox of the bed or alternatively the two materials may be allowed to mix within the bed itself by admitting them through separate distributors. In the former case an explosive gas–air mixture is produced in the windbox, although such pre-mixed gas–air mixtures have been burned successfully in the laboratory and on an industrial scale. The normal procedure however is for the fuel and air to be admitted separately to the bed–in the case of fuel oils from nozzles or 'guns' located either in the main gas distributor or in the bed wall. Mixing is then brought about by the hydrodynamic forces operating within the bed. Flammable gas–air mixtures flowing interstitially will burn smoothly at bed temperatures in excess of about 700°C, but such a mixture flowing in the bubble phase will tend to burn explosively and to generate the noise which is a characteristic feature of fluidized bed gas combustion. A fuel-rich bubble that rises through the bed without coalescing en route with an air bubble big enough to produce a flammable mixture will release gas into the freeboard region, where it will mix readily with the excess air present and burn as an unstabilized flame above the bed surface. It is thus essential if all the fuel is to be burned within the bed for the gas to be well mixed with air before it breaks the bed surface; unfortunately gas mixing in fluidized beds is generally poor and so careful attention must be paid to the design of fuel and air distributors if the escape of unburned gas into the freeboard is to be avoided. (But see the remarks made earlier in connection with the plume model of Park *et al.*[25]).

A number of experimental studies of fluid bed oil and gas combustion have been published. Elliott and Virr[67] described measurements made with a laboratory-scale combustor in which a premixed town-gas/air mixture was burned in a shallow sand bed fitted with a ceramic plate distributor. Bed temperatures in the range 800–1200°C were obtained with combustion intensities of around 10 MW/m^3.

Work on a bigger scale in which liquified petroleum gas (LPG) was burned in a bed of sand has been described by Singh *et al.*[68]. Here the fuel and air were admitted separately to the bed through specially designed distributors which enabled gases to be injected at different locations in the bed and which were mechanically stable at temperatures of up to 850°C. Each distributor consisted of two coaxial pipes, the centre pipe having a single row of orifices and the outer pipe a double row. Two beds were used, one cylindrical in shape with a diameter of 360 mm and the other rectangular with dimensions 910 mm × 610 mm; in both cases the LPG distributor was situated above the air distributor.

Singh *et al.* observed freeboard combustion under a variety of conditions. With the rectangular bed and a sand depth of 500 mm a stoichiometric mixture flowing at 355 m^3/hr burned completely within the bed. On reducing the flow to 152 m^3/hr flames appeared above the bed but were eliminated by increasing the air flow while maintaining the LPG flow constant. These observations are consistent with a reduction in fuel-and air-bubble coalescence rate at the lower gas velocities. From the results of gas analyses made at three bed heights in the circular bed with a constant air flow and a varying fuel flow it was apparent that with a bed height of 90 mm and for LPG flows in excess of 1.1 m^3/hr there was insufficient oxygen to prevent the breakthrough of unburned gas. The appearance of combustibles above the bed was suppressed however when the bed height was increased to 225 mm, providing further evidence of the effect of bubble coalescence in deeper beds. It is noteworthy that Elliott and Virr achieved 100 per cent combustion efficiency in beds only 12 mm deep with premixed fuel–air mixtures. An unusual observation made by Singh *et al.* with their rectangular bed was that appreciable lateral temperature gradients could be established within the bed by variation of the LPG flowrate. It was found possible for example to maintain one end of the bed at 850°C and the other end at 700°C by virtue of the rather poor lateral circulation of solid particles under normal operating conditions. Measurements carried out with solid tracers showed that whereas uniform mixing in the vertical direction could be achieved in 75 s at a gas velocity of three times the incipient fluidizing value, mixing over the length of the bed was complete only after 1.5 hr.

Reh[69] has given a brief description of the industrial-scale combustion of refinery waste sludges containing about 12 per cent oil, 70 per cent water and 18 per cent ash. This was carried out in beds with internal diameters of 1.4 m containing sand fluidized with air. The oil sludge was injected through nozzles situated in the bed wall and provision was made for similarly introducing fuel gases to maintain bed temperature at the required value. A considerable amount of afterburning of the gaseous fuel occurred under most conditions; it was particularly pronounced when oil sludges with very low calorific values were treated, sometimes causing damage to the heat exchangers situated downstream of the bed.

A paper by Stubbington and Davidson[70] reported an experimental study of the combustion of both kerosine and propane injected from a single entry point into an air-fluidized bed of sand at 940°C. They determined the amount of fuel burnt in the bed from heat balances and investigated the influence on bed-combustion efficiency of particle size, bed height and fluidizing velocity. They assumed the fuel entered the bed as a stream of bubbles and showed that its subsequent combustion could be divided into two stages; (i) in the region between successive bubbles close to the injection point, and (ii) on a reaction 'surface' within the bed to which fuel and oxygen had diffused. The latter gave rise to a laminar diffusion flame within the emulsion phase of the bed. They also showed that overbed burning resulted mainly from a bypassing mechanism in which fuel-rich bubbles rise through the bed without contacting oxygen; this was particularly important in deep beds of small diameter particles where the gas bubbles are surrounded by clouds.

4.6 Desulphurization

Sulphur compounds present in coal (thiols, thiophens, sulphides, etc.) are oxidized during its combustion to sulphur dioxide. In the presence of a basic solid such as calcium oxide, produced *in situ* by calcination of limestone or dolomite: $CaCO_3 = CaO + CO_2$, the SO_2 will react to form a solid product, calcium sulphate, and is thereby removed from the off gases.

Under the usual conditions of operation of a fluid-bed combustor, two reaction sequences leading to calcium sulphate are thermodynamically feasible:

$$\left.\begin{array}{lcl} CaO + SO_2 & \rightleftharpoons & CaSO_3 \\ CaSO_3 + \tfrac{1}{2}O_2 & = & CaSO_4 \end{array}\right\} \text{route 1}$$

$$\left.\begin{array}{lcl} SO_2 + \tfrac{1}{2}O_2 & \rightleftharpoons & SO_3 \\ CaO + SO_3 & = & CaSO_4 \end{array}\right\} \text{route 2}$$

The second route is followed only if SO_3 producing catalysts, such as heavy metal salts, are present in the bed.

A problem with using limestone in this way however is that since the molar volume of $CaSO_4$ (52.2 cm^3/mol) is greater than that of $CaCO_3$ (36.9 cm^3/mol) and since no particle expansion occurs during the sulphation, the maximum degree of limestone conversion possible to realize is about 50 per cent[71]. Moreover, since material at the pore entrances sulphates before that in the internal porous structure of the calcined limestone, these entrances become blocked with $CaSO_4$ and so the practical maximum conversion is normally very much less than 50 per cent (*see* Section 4.6.2).

Extensive studies of coal-burning systems operating at atmospheric pressure have been reported by the National Coal Board[72], the Argonne National Laboratory[73], Westinghouse[74], and Exxon[75]. In addition, several groups have investigated coal desulphurization at elevated pressures[2,6,8], and fuel-oil desulphurization has also received attention[76]. In general these studies have been directed towards obtaining correlations and data that may be applied to the engineering design of full-scale plant, and they have attempted to quantify the design parameters over a wide range of operating conditions.

4.6.1 Design parameters

(i) *Ca/S mole ratio*

Since the solid calcium sulphate tends to block the pores of the stone particles and thus to limit the extent of their conversion it is necessary for high levels of SO_2 retention to operate with quantities of sorbent in excess of the stoichiometric amount. This excess is normally expressed in terms of the ratio of calcium in the stone to sulphur in the fuel and, as is to be expected, the efficiency of SO_2 removal increases with this ratio. Highley[77] has given the empirical relationship:

$$R = 100[1 - \exp(-M\beta)] \qquad (4.15)$$

for the percentage reduction in SO_2 emission (R) in terms of Ca/S mole ratio β, and a constant M, whose value depends on coal, additive and operating conditions.

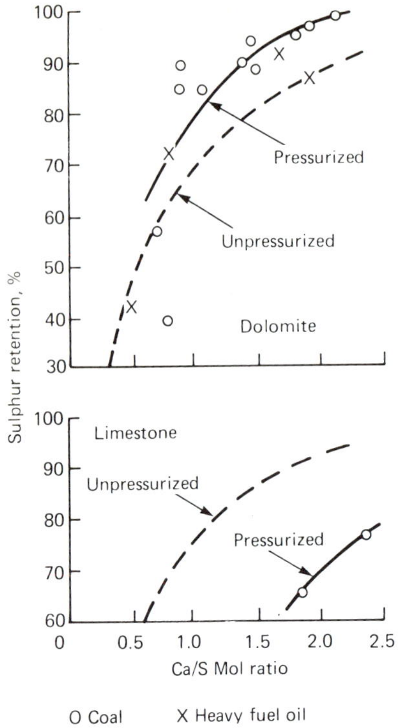

Figure 4.13 Effect of Ca/S mol ratio on sulphur retention (From Roberts et al.[6]*)*

The effect of Ca/S mole ratio at elevated pressures (350–600 kN/m^2 and 800°C) is shown in *Figure 4.13* for both limestone and dolomite sorbents[6]. The poorer performance of limestone at the higher pressures is due to the displacement of the equilibrium of calcination to the left:

$$CaCO_3 \rightleftharpoons CaO + CO_2$$

Thus the open pore structure of calcined lime is less available for the sulphation reaction, which is then confined to the outer layers of the particles. Since the half-calcination of dolomite:

$$MgCO_3.CaCO_3 \rightleftharpoons MgO.CaCO_3 + CO_2$$

occurs at lower temperatures, its absorbent properties should be relatively unaffected by an increase in the partial pressure of CO_2, and in practice dolomite performs better at higher pressures. It has been shown however[78] that MgO reacts only slowly with SO_2 at temperatures between 540° and 980°C, and it is suggested that the high reactivity of dolomite is due to the small grain size of its CaO component which is formed independently of MgO during calcination.

(ii) *Bed temperature*
It is well established that under atmospheric pressure conditions the most favourable temperature for the sulphation reaction of both lime and dolomite is in the region of 800–850°C[6,12,76,77], the retention of SO_2 falling quite sharply with increase in temperature above this (*Figure 4.14*). No such effect has been observed however in operation at elevated pressures where the SO_2 retention increases with temperature up to at least 900°C[6]. The reasons for these observations are not well understood although one possible explanation of the optimum temperature effect as being due to the

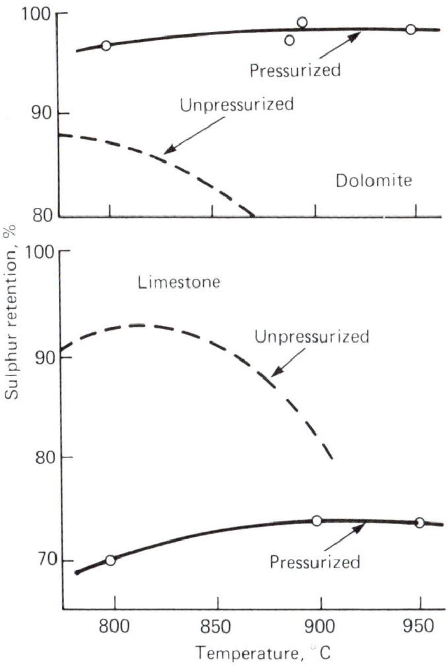

Figure 4.14 Effect of temperature on sulphur retention (From Roberts et al.[6])

reversible decomposition of calcium sulphate is ruled out on thermodynamic grounds[6,77]. A recent study has shown that the effect could be due to regeneration of SO_2 by CO at the higher temperatures[79]:

$$CaSO_4 + CO \rightarrow CaO + CO_2 + SO_2.$$

(iii) *Choice of sorbent*
The criteria for choosing a suitable sorbent for SO_2 in a fluid bed combustor have been listed by O'Neill and Keairns[80] as:
(a) the acceptor properties of the stone for SO_2
(b) attrition resistance
(c) trace and minor element emission characteristics
(d) regeneration characteristics
(e) suitability of the spent sorbent for final processing and disposal
(f) economic availability

The degree to which these are met varies not only between limestone and dolomite, the two materials most extensively investigated, but also among limestones from different sources. A determining factor for several of the above criteria is the type of pore structure produced on calcination. Laboratory tests by the NCB showed that SO_2 absorption with US limestone No. 1359 (from Stephens City, Virginia) occurred within pores of diameter less than 0.3 μm, but that, as sulphation progressed, these pores became blocked with sulphate. The plant performance of the same material was however better than predicted from the laboratory experiments, due probably to particle attrition in the fluidized bed causing fresh surface to be exposed for reaction. Similar results have been reported by the Argonne group[73] and by Ehrlich[12].

Experiments using a 15 cm diameter pressurized fluid-bed combustor with Tymochtee dolomite and limestone No. 1359 were described by Vogel *et al.*[8]. Runs were carried out at a pressure of 8 bar and at temperatures of 900 and 950°C, but it was only at the higher temperature that the $CaCO_3$ in the dolomite was calcined, and as a consequence the material showed better sulphur retention properties than at 900°C. The sorption properties of the limestone were unaffected by bed temperature at a fixed Ca/S ratio and were inferior to those of the dolomite under all conditions.

A laboratory study of a number of limestones and dolomites using thermogravimetric analysis[80] has indicated ways in which sorbent screening may be carried out, but the point is made that tests under actual operating conditions in a fluid bed combustor are still necessary before a definitive choice of material can be made.

(iv) *Sorbent regeneration*

Because of the enormous quantities of stone that would be needed if it were to be used on a once-through basis only, the large-scale application of fluid-bed combustion will require the sorbent to be regenerated and recycled continuously. Two regeneration processes have been suggested[81]. The first is a one-step process in which the sulphated stone is reduced with hydrogen or carbon monoxide in a separate fluidized bed reactor at 1100°C:

$$CaSO_4 + H_2 \rightleftharpoons CaO + H_2O + SO_2$$

Under these conditions a side reaction also occurs leading to the formation of calcium sulphide. Regenerated lime is returned to the fluid bed combustor while the SO_2 is sent to a sulphur recovery plant for conversion to sulphuric acid or elemental sulphur. The second suggested process involves a two-step reduction via calcium sulphide at 800°C and 10 bar:

Step 1: $CaSO_4 + 4H_2 \rightleftharpoons CaS + 4H_2O$
Step 2: $CaS + H_2O + CO_2 = CaCO_3 + H_2S$
Step 2 requires a temperature of about 680°C and a pressure of 10 bar.

Although the feasibility of the one-step reduction to CaO and SO_2 has been demonstrated on the laboratory scale using both propane[2] and fuel oil[76,82] as the reducing agent, the integration of stone regeneration into a large-scale fluid-bed combustion unit has not yet been accomplished.

4.6.2 Sulphation models

As was mentioned earlier, the reaction between calcined limestone and gaseous sulphur dioxide results in the formation of solid calcium sulphate which, owing to its having a greater molar volume than the original calcium carbonate, blocks the pores of the stone and prevents its complete utilization. In the initial stages of exposure to the gas, reaction occurs almost entirely on the outer surface of the particles, but with increasing time a reaction zone moves inwards towards the centre and a sulphate concentration gradient develops throughout the particle. Pore blockage proceeds inwards gradually from the outer regions with both the rate of reaction and the total degree of conversion depending strongly on the porosity of the original material[83]. In most cases the rate of sulphation decreases exponentially with time[84] and the kinetics are first order in SO_2[85,86] and zero order in oxygen at the high excess air levels normally employed.

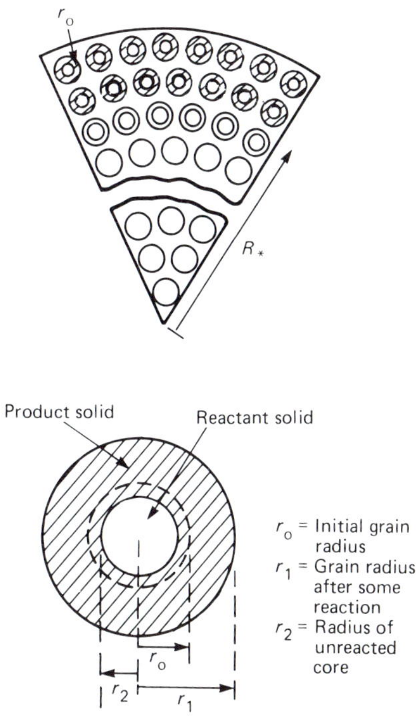

Figure 4.15 Schematic representation of changing size grain model (From Georgakis et al.[93])

Although it may seem reasonable to model the process on the basis of the 'shrinking-core' model of gas-solid reactions[34], the porous structure of calcined limestone particles makes the so-called 'grain theory', introduced by Szekely and Evans[87,88], more appropriate to the sulphation reaction. This theory assumes the particles to be composed of grains of reactive solid, separated by pores through which reactive gases diffuse (*Figure 4.15*). As reaction proceeds a surface layer of product builds up on the grains and diffusion of reactant gas as increasingly inhibited. In cases where reaction leads to an increase in the molar volume of the solid, a point is reached where the resistance to diffusion is so great that gas cannot penetrate to the unreacted material beneath the product layer and reaction effectively ceases.

The application of this model to the limestone sulphation reaction has been described by Pigford and Sliger[89], Hartman and Coughlin[90], Ramarchandran and Smith[91] and by Georgakis *et al.*[84,92,93], and in most cases good agreement between theory and experiment has been found. A disadvantage of this approach however is that the model equations are generally non-linear and it is difficult to incorporate them into a hydrodynamic model of a fluidized-bed combustor without having to resort to prohibitively lengthy computer calculations. As a result, a number of simpler, semi-empirical models have been developed by Lee *et al.*[84], Fieldes and Davidson[94] and by Zheng *et al.*[95]. The latter workers used the analogy between limestone sulphation and the deactivation of catalyst particles by coke formation, both of which can be described in terms of an exponential decrease in available surface with time:

$$S = S_o \exp(-k_d t) \tag{4.16}$$

Here S_o is the initial surface area of reactant particles and k_d is a 'deactivation' rate constant. The overall rate of reaction, R, may be expressed for a completely mixed system as:

$$R = k_s C_A S \tag{4.17}$$

where k_s is a surface reaction rate constant, C_A is the concentration of SO_2 in the exit gases and S is the available surface of calcium oxide. Combining these two equations with a mass balance on SO_2, assuming the fluid bed to be a completely mixed system, an expression is obtained which gives the outlet concentration of SO_2 in terms of two time-invariant rate constants, k_d and k_s, the mass of limestone added in a batch, M_b, and the air flowrate Q.

$$\ln\left[\left(\frac{C_{Ao}}{C_A}\right) - 1\right] = \ln\left(\frac{6k_s M_b}{\rho_s \bar{d}_p Q}\right) - k_d t \tag{4.18}$$

ρ_s and $\bar{d}_p$ are the density and mean particle size respectively of the limestone particles. Both k_d and k_s contain chemical and mass transfer terms but they can be determined empirically for a wide range of operating conditions by relatively straightforward batch experiments. Equation 4.18 may then be linked to a model of a fluid-bed combustor to give an expression for the fractional SO_2 retention, η:

$$\eta = 1 - \frac{1}{1 + K_\beta} \tag{4.19}$$

Here K is a function of limestone type and operating conditions and β is the calcium-to-sulphur mole ratio. Zheng *et al.* used equation 4.19 to obtain predictions for continuous operation and found good agreement with experimental values from a small-scale laboratory unit. It is worth noting the difference between equation 4.19 and that obtained by Highley (equation 4.15) for larger-scale equipment.

4.7 NO_x emission

It is generally accepted that the oxides of nitrogen formed during fluid-bed combustion originate predominantly from the nitrogen compounds contained in the fuel. Results obtained at BCURA and reported by Roberts *et al.*[6] are summarized in *Table 4.3*. The data refer to combustion measurements made over a range of temperatures from 800 to 950°C with 20 per cent excess air under which conditions 90–95 per cent of the NO_x consisted of nitric oxide NO, the remainder being NO_2. Ehrlich has given the following correlation for the concentration of NO_x in the off-gases from coals containing 1.1–2.2 per cent nitrogen at bed temperatures between 785 and 980°C[12]:

Table 4.3 Concentration of nitrogen oxides in gas from fluidized bed combustors (from Roberts *et al.*[6])

Fuel	*Gas*	*Oil*	*Coal*
Combined nitrogen (mass %)	Nil	0.2	1.0–1.5
	NO_x *content of flue gas* (ppm)		
Unpressurized	20–30	130	300–600
Pressurized (350–600 kN/m^2)	Not determined	110	80–130

$$NO_x \text{ (vol. ppm)} = 0.28 \times 10^6(1 + 1.7N)K \qquad (4.20)$$

where K is the equilibrium constant for the reaction $N_2 + O_2 = 2NO$, and N is the weight percentage of nitrogen in the fuel (dry basis). This relationship is consistent with results reported by Gibbs *et al.*[7] from experiments in which axial and radial concentration profiles of NO were measured in a coal burning bed (*Figure 4.16*). These results show that NO is formed preferentially near the coal inlet from which the authors conclude that the volatile components of the coal contribute to most of the NO formation above 750°C and that this gas is swept out of the bed in the bubble phase. Further experiments showed that NO_x emissions could be reduced by burning a mixture of coal and char in the bed, the nitric oxide being reduced to nitrogen by reaction with the char. Gibbs *et al.* found a relatively weak dependence between NO_x emission and stoichiometric ratios, but showed that above 650°C the emission in the flue gases increased linearly with temperature at a rate of 2.6 ppm/°C.

As is indicated in *Table 4.3*, the level of NO_x emission is reduced considerably by operation at elevated pressures although, unlike the effect

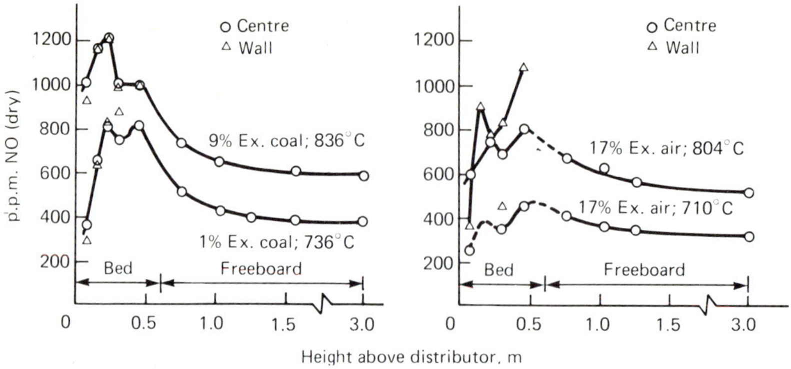

Figure 4.16 Axial and radial NO concentration profiles (From Gibbs et al.[7])

at atmospheric pressure, increasing excess air leads to an increase in NO_x[2,8]. However results reported by Exxon[2] from two fluid bed units operated at 800 and 900 kN/m^2 respectively, show the emissions for a likely operating excess air ratio of 15 per cent to be well within the current US EPA standard.

4.8 Coal gasification

Processes for converting coal into gas have a history stretching back to the early nineteenth century. The product of these processes was town gas, a blend of the volatile constituents of coal with water gas formed by blowing steam through the devolatilized char. Coke was a by-product and its market value was often a crucial factor in determining the profitability of the process. Coal gas plants were generally built on a very big scale and produced a gas containing a wide range of impurities such as hydrogen sulphide, organic sulphur compounds and unsaturated hydrocarbons. Town gas had a calorific value in the range 20–24 MJ/m^3 (500–600 Btu/ft^3) and, as a result of being about 50 per cent hydrogen, was characterized by a high flame speed. Gasification was normally carried out under atmospheric pressure and little if any pretreatment was given to the coal feed.

In the decade or so before the 1939–45 War a number of new processes were developed, mostly in Germany, with the objective of gasifying all the organic matter in coal and leaving ash as the only solid residue. The coal particles which were prepared in a pretreatment step were kept in motion during reaction by being fluidized as in the Winkler process, dust injected as in the Koppers–Totzec process or suspended in a molten slag as in the Rummel process. The success of these 'second generation' gasifiers was however limited; they were not always able to deal with a wide range of coals nor to prevent the loss of ungasified carbon in the elutriated ash or molten slag.

One of the more successful of these processes was that developed by the

Lurgi Gesellschaft, versions of which have been installed in the UK, Australia and South Africa as well as in Germany itself, while the plant that has been operated for many years by Sasol in South Africa as a source of synthesis gas for Fischer–Tropsch gasoline is reported to be undergoing considerable expansion. The Lurgi gasifier operates under pressure (20–30 bar) and uses a steam/oxygen mixture as the gasifying medium. The feedstock is lump coal which is admitted to the reactor via a pressurized hopper and kept in motion during reaction by means of a rotating grate through which the ash is discharged. The steam/oxygen ratio is kept high in order to prevent the ash from fusing and causing damage to the grate, although recent developments to the process have aimed at increasing the temperature of operation and discharging the ash as a liquid slag. High pressure operation is advantageous in that it not only allows a higher throughput but also causes the yield of methane to be increased by promoting hydrogenation of the coal carbon: $C + 2H_2 = CH_4$.

Nevertheless, the methane content of the crude gas is only of the order of 4.7 per cent while hydrogen and carbon monoxide account for 20.1 and 9.2 per cent respectively. It is thus necessary if the product is to be used as a natural gas substitute to include a methanation stage in the normal gas treatment train. In this way the calorific value of the Lurgi gas may be raised from the 12.0 MJ/m^3 (*c*. 300 Btu/ft^3) of the crude product to the required level of around 38.0 MJ/m^3 (*c* 950 Btu/ft^3). Two Lurgi plants were built in the UK in the early 1960s for the production of town gas but they were never able to compete economically with the oil-reforming processes (such as the ICI '500' process) developed at about the same time. The plant situated at Westfield near Edinburgh has however been used for a development programme sponsored by a number of US firms headed by Conoco, the work being directed towards evaluating the 'slagging' gasifier for use with strongly caking US coals and also to demonstrate the large-scale methanation of Lurgi gas[96].

The first industrial process to make use of the idea of a gas–solid fluidized bed was the Winkler coal-gas process (*Figure 4.17*); it was patented in 1922 and began operation at the BASF plant at Leuna, Germany in 1926. Since then some sixteen Winkler generators have been built in various parts of the world and have been used for the production of low calorific value gas (11 MJ/m^3 or 275 Btu/ft^3) suitable as an industrial fuel or as a synthesis gas for methanol or ammonia production; it is believed that none are operating at the present time. In the reactor section of the process, dried coal, sized to 0–10 mm is screw-fed from the storage bunker into the gasifier where it is fluidized with air (and/or oxygen) and steam at atmospheric pressure and at a temperature, depending on the type of coal used, in the range 1100–1300 K. At these high temperatures all tars and heavy hydrocarbons are reacted.

During reaction, ash particles undergo segregation, the larger agglomerates, representing about 30 per cent of the total, settling to the bottom of the bed and being removed by a screw discharger; the lighter particles are

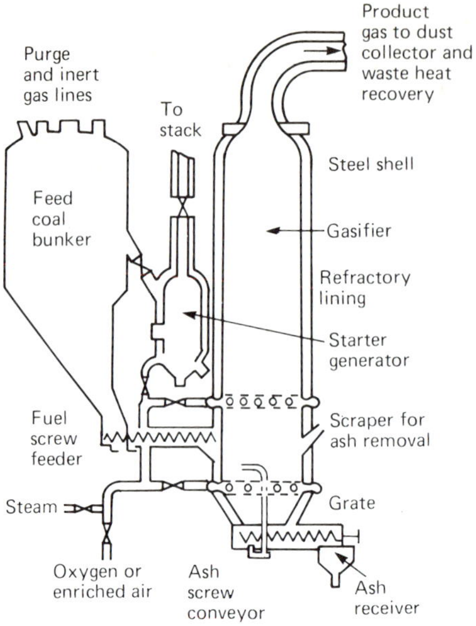

Figure 4.17 Winkler gasifier (From Cooke and Robson[97])

elutriated with the product gas and unreacted coal fines. These unreacted coal particles are gasified by a secondary steam/air mixture injected into the freeboard region, but the efficiency of conversion is not high and some 20 per cent of the carbon fed to the generator can be lost with the fly ash[98]. This fact, coupled with the low throughput resulting from its low pressure operation, makes the Winkler process inefficient by modern standards although process improvements are obviously possible.

Over the last two decades, in several of the technologically more advanced areas of the world, pipeline gas manufactured from coal or oil has been replaced by natural gas consisting mainly of methane. Owing to its cheapness, cleanness and general convenience of use, the demand for natural gas from both domestic and industrial consumers has increased at a rapid rate, so much so that by the early 1970s it had become clear that within a few years supplies from known indigenous reserves would be insufficient to meet requirements. The problem was particularly acute in the USA where estimates predicted a widening gap between supply and demand, a likely deficiency of 0.368×10^{12} m^3 in 1980 increasing to 0.850×10^{12} m^3 by 1995[99], even making allowance for the newly discovered sources in Alaska. It has become generally accepted that the only realistic solution is a return to the use of a manufactured gas based on coal, reserves of which are estimated to be sufficient to satisfy current levels of demand for some hundreds of years. This supplemental natural gas (SNG) must of course be fully compatible and interchangeable with natural gas itself, i.e. it must contain of the order of 90 per cent methane. A successful process must produce SNG at a cost not much different from that of natural gas and must be able to meet the increasingly strict standards relating to environmental pollution. Despite the enormous problems, a number of organizations have embarked on intensive programmes to develop a viable coal gasification process. Commercialization is planned for the 1980s and at the time of writing the only results that have appeared are from pilot plant studies. Nevertheless, details of a number of promising new gasification methods are beginning to emerge and it is clear that the majority of them will be based on the extensive use of fluid-bed technology.

4.8.1 The Synthol process

At their Sasolburg plant in South Africa the Sasol company have since 1955 operated large-scale units for the production of a range of coal-derived gases and liquids. The technology is based on a combination of the Lurgi process, producing a synthesis gas mixture of carbon monoxide and hydrogen, and the Fischer–Tropsch process, which catalytically converts the synthesis gas to the required products–mainly light oils and gasoline. A total of thirteen Lurgi gasifiers are currently installed, each of some 4 m internal diameter and producing in total an average of around 8.2×10^6 m^3/day of raw gas. The Lurgi reactors supply gas to two Fischer–Tropsch variants, the Arge process and the Synthol process. The Arge plant consists of five packed-bed reactors containing a pelletized precipitated iron catalyst; each reactor can produce some 18 000 tonnes/year of hydrocarbons.

The Synthol plant has three transported-bed reactors, each with a capacity of 65 000 tonnes/year of hydrocarbons. Synthol catalyst is a promoted iron powder derived from the millscale produced in steel making. In the process a mixture of fresh and recycle gas at 160°C and 21.5 bar enters through a pipe at the base of the unit and comes into contact with catalyst. The catalyst flows out of the standpipe via a slide valve and is entrained by reactant gas in the same way as is the regenerated catalyst in a fluid catalytic cracking (FCC) reactor. The mixture of gas and entrained solid rapidly equilibrates at about 315°C but then begins to heat up as the exothermic synthesis reactions start to occur:

$$
\begin{aligned}
nCO + (2n + 1)\,H_2 &= C_nH_{2n+2} + nH_2O \\
nCO + 2nH_2 &= Cn\,H_{2n} + nH_2O \\
CO + H_2O &\rightleftharpoons CO_2 + H_2
\end{aligned}
$$

Heat is removed in the coolers situated in the expanded section of the transport pipe and the temperature reaches a maximum of 340°C in the hopper above the standpipe. Gas and catalyst are separated in the hopper, the gas leaving the reactor via the cyclones, to be scrubbed, separated and purified, and the catalyst returning to complete its cycle via the standpipe. The rate of catalyst circulation is controlled by means of slide valves in the standpipe, a procedure also adopted in FCC systems. A batch of catalyst has a useful life of about 50 days, after which its selectivity has deteriorated away from gasoline towards the lighter hydrocarbons, and it is replaced.

The Synthol capacity at Sasolburg is currently being increased by an order of magnitude and when complete the plant will be able to satisfy the gasoline demand of the entire South African economy. The economic viability of the Lurgi–Synthol route to gasoline in any other country at the present time remains in doubt however.

References

1 SKINNER, D.G., *The Fluidised Combustion of Coal*, Mills and Boon, London (1971)
2 HOKE, R.C. and BERTRAND, R.R., *Inst. Fuel Symp. Ser.*, No. 1, Vol. 1, D5 (1975)
3 LOCKE, H.B., LUNN, H.G. and ROBERTS, A.G., *Proc. 3rd Int. Conf. Fluidized Bed Combustion*, Paper 5, US-EPA (1972)
4 ELLIOT, D.E., *Proc. 2nd Int. Conf. Fluidized Bed Combustion*, Paper 0–1 US-EPA (1970)
5 HOY, H.R. and ROBERTS, A.G., *A.I.Ch.E. Symp. Ser.*, **68**(126), 225 (1972)
6 ROBERTS, A.G., STANTAN, J.E., WILKINS, D.M., BEACHAM, B. and HOY, H.R., *Inst. Fuel Symp. Ser.*, No. 1, Vol.1, D4 (1975)
7 GIBBS, B.M., PEREIRA, F.J. and BEER, J.M., *Inst. Fuel Symp. Ser.*, No. 1, Vol.1, D6 (1975)
8 VOGEL, G.I., SWIFT, W.M., MONTAGNA, J.C., LENC, J.F. and JONKE, A.A., *Inst. Fuel Symp. Ser.*, No.1, Vol.1, D3 (1975)
9 BISHOP, J.W., US Patent 3508506 (1968)
10 EHRLICH, S., ROBISON, E.B., GORDON, J.S. and BISHOP, J.W., *A.I.Ch.E. Symp. Ser.*, **68**(126), 231 (1972)
11 BISHOP, J.W., *Proc. 2nd Int. Conf. Fluidized Bed Combustion*, Paper 4–4, US-EPA (1970)
12 EHRLICH, S., *Inst. Fuel Symp. Ser.*, No. 1, Vol.1, C4 (1975)
13 KEAIRNS, D.L., HAMM, J.R. and ARCHER, D.H., *A.I.Ch.E. Symp. Ser.*, **68**(126), 267 (1972)
14 GODEL, A.A., *Rev. Gen. de Thermique*, **5**, 349 (1966)
15 SVOBODA, J.J., *Proc. 2nd Int. Conf. Fluidized Bed Combustion*, Paper 1–6 US-EPA (1970)
16 TU, C.M., DAVIS, H. and HOTTEL, H.C., *Ind. Engng. Chem.*, **26** 749 (1934)
17 NUSSELT, W.Z., *Deut. Ing.*, **68**, 124 (1924)
18 BURKE, S.P. and SCHUMANN, T.E.W., *Proc. 3rd Int. Conf. Bituminous Coal*, **2**, 485 (1931)
19 SPALDING, D.B., *Some Fundamentals of Combustion*, Gas Turbine Series, Vol. 2, Butterworths, London (1955)
20 FIELD, M.A., GILL, D.W., MORGAN, B.B. and HAWKSLEY, P.G.W., *Combustion of Pulverised Coal*, BCURA, Leatherhead, UK (1967)
21 MULCAHY, M.F.R. and SMITH, I.W., *Rev. Pure and Appl. Chem.*, **19**, 81 (1969)
22 SMITH, I.W., *Comb. and Flame*, **17**, 303 (1971)
23 CARAM, H.S. and AMUNDSON, N.R., *Ind. Engng. Chem. Funds.*, **16**, 171 (1977)
24 AVEDESIAN, M.M. and DAVIDSON, J.F., *Trans. I. Chem. Engrs*, **51**, 121 (1973)
25 PARK, D. LEVENSPIEL, O. and FITZGERALD, T.J., *Chem. Engng. Sci.*, **35**, 295 (1980)
26 PYLE, D.L., *Inst. Fuel Symp. Ser.*, No. 1, Vol. 2, 17 (1975)
27 BORGHI, G., SAROFIM, A.F. and BEER, J.M., *A.I.Ch.E.* 70th Annual Meeting, New York (1977)
28 ATIMTAY, A., in *Fluidization*, Eds. Grace, J.R. and Matsen, J.M., 159 Plenum Press, New York (1980)
29 YATES, J.G., MACGILLIVRAY, H.J. and CHEESMAN, D.J., *Chem. Engng. Sci.*, **35**, 2361 (1980)
30 ORMSTON, D., ROBINSON, E. and BUCKLE, D., *Inst. Energy Symp. Ser.*, No. 4, Vol. 1, 2A-1 (1980)
31 HOUGEN, O.A. and WATSON, K.M., *Chemical Process Principles*, Part 3, Wiley, New York (1947)

32 CAMPBELL, E.K. and DAVIDSON, J.F., *Inst. Fuel Symp. Ser.*, No. 1, Vol. 1, A2 (1975)
33 YAGI, S. and KUNII, D., *5th Int. Symp. Combustion*, 231, Reinhold, New York (1955)
34 LEVENSPIEL, O., *Chemical Reaction Engineering*, 2nd Edn., Wiley, New York (1972)
35 TOMECZEK, J., *Trans. I. Chem. Engrs*, **57**, 215 (1979)
36 BASU, P., BROUGHTON, J. and ELLIOTT, D.E., *Inst. Fuel Symp. Ser.*, No. 1, Vol. 1, A3 (1975)
37 CHAKRABORTY, R.K. and HOWARD, J.R., *J. Inst. Fuel*, 220 (1978)
38 ROSS, I.B. and DAVIDSON, J.F., *Trans. I. Chem. Engrs*, **60**, 108 (1982)
39 LA NAUZE, R.D., *Inst. Energy Symp. Ser.*, No. 4, Vol. 2, KA-1 (1980)
40 ROSS, I.B. and DAVIDSON, J.F., *Trans. I. Chem. Engrs*, **57**, 216 (1979)
41 BHATTACHARYYA, D. and PEI, D.C.T., *Ind. Eng. Chem. Funds.*, **13**, 199 (1974)
42 CHAKRABORTY, R.K. and HOWARD, J.R., *A.I.Ch.E. 86th National Meeting*, Houston, Session 57, Paper D (1979)
43 CHAKRABORTY, R.K. and HOWARD, J.R., in *Fluidization*, Eds. Grace, J.R. and Matsen, J.M., 167, Plenum Press, New York (1980)
44 YATES, J.G. and WALKER, P.R., in *Fluidization*, Eds. Davidson, J.F. and Keairns, D.L., 241, Cambridge University Press (1978)
45 ZABRODSKY, S.S. and PARNAS, A.L., *Inst. Fuel Symp. Ser.*, No. 1, Vol. 1, B5 (1975)
46 WHELLOCK, J.G., *Chem. Engineer* (Birmingham University), **23**, 18 (1972)
47 ROSCOE, J.C., WITOWSKI, A.R. and HARRISON, D., *Trans. I. Chem. Engrs*, **58**, 69 (1980)
48 ROSS, I.B., PATEL, M.S. and DAVIDSON, J.F., *Trans. I. Chem. Engrs.*, **59**, 83 (1981)
49 GIBBS, B.M., *Inst. Fuel Symp. Ser.*, No. 1, Vol. 1, A5 (1975)
50 GORDON, A.L. and AMUNDSON, N.R., *Chem. Engng. Sci.*, **31**, 1163 (1976)
51 GORDON, A.L., CARAM, H.S. and AMUNDSON, N.R., *Chem. Engng. Sci.*, **33**, 713 (1978)
52 CHEN, T.P. and SAXENA, S., *Fuel*, **56**, 401 (1977)
53 HORIO, M., MORI, S. and MUCHI, I., *Proc. 5th Int. Conf. Fluidized Bed Combustion*, **2**, 605 (1977)
54 HORIO, M. and WEN, C.Y., *A.I.Ch.E. Symp. Ser.*, **74**(176) 101 (1978)
55 RAJAN, R., KRISHNAN, R. and WEN, C.Y., *A.I.Ch.E. Symp. Ser.*, **74**(176), 112 (1978)
56 BARON, R.E., BEER, J.M., BORGHI, G., HODGES, J.L. and SAROFIM, A.F., *Proc. 5th Int. Conf. Fluidized Bed Combustion*, **3**, 437 (1977)
57 BARON, R.E., HODGES, J.L. and SAROFIM, A.F., *A.I.Ch.E. Symp. Ser.*, **74**(176), 120 (1978)
58 FAN, L.T., TOJO, K. and CHANG, C.C., *Ind. Engng. Chem. Funds.*, **18**, 333 (1979)
59 DONSI, G., MASSIMILLA, M.M., RUSSO, G. and STECCONI, P., *Combust. Sci. Technol.*, **21**, 25 (1979)
60 BADZIOCH, Z., *BCURA Month. Bul!.*, **25**, 285 (1961)
61 HIGHLEY, J. and MERRICK, D., *A.I.Ch.E. Symp. Ser.*, **67**(116), 219 (1971)
62 GIBBS, B.M. and BEER, J.M., *I. Chem. Engrs. Symp. Ser.*, No. 43, Paper 21 (1975)
63 WATERS, P.L., *Inst. Fuel Symp. Ser.*, No. 1, Vol. 1, C6 (1975)
64 GIBBS, B.M. and HEDLEY, A.B., in *Fluidization*, Eds. Davidson, J.F. and Keairns, D.L., 235, Cambridge University Press (1978)
65 DONSI, G., MASSIMILLA, L. and MICCIO, M., *Comb. and Flame*, **41**, 57 (1981)
66 BEER, J.M., MASSIMILLA, L. and SAROFIM, A.F., *Inst. Energy Symp. Ser.*, No. 4, Vol. 1, 4–5 (1980)
67 ELLIOTT, D.E. and VIRR, M.J., *Proc. 3rd Int. Conf. Fluidized Bed Combustion*, Paper 4–1 US-EPA (1972)
68 SINGH, B., RIGBY, G.R. and CALLCOTT, T.G., *Inst. Fuel Symp. Ser.*, No. 1, Vol. 1, C5 (1975)
69 REH, L., *Proc. 2nd Int. Conf. Fluidized Bed Combustion*, Paper 5–6, US-EPA (1970)
70 STUBBINGTON, J.F. and DAVIDSON, J.F., *A.I.Ch.E. Jl*, **27**, 59 (1981)
71 HARTMAN, M. and COUGHLIN, R.W., *Ind. Eng. Chem. Proc. Des. Dev.*, **13**, 248 (1974)
72 *Reduction of Atmospheric Pollution*, Vol. 3, National Coal Board, London (1971)
73 VOGEL, G.J. *et al.*, Argonne National Lab. Reports, EPA-650/2-74-104 (1974), EPA-650/2-74-057 (1973)

74 KEARINS, D.L. *et al.*, Westinghouse Research Lab. Report, EPA-650/2-73-048a-c (1973)
75 HOKE, R.C. *et al.*, Exxon Research and Engineering Co. Report, EPA-650/2-74-001 (1973)
76 MOSS, G., *Inst. Fuel Symp. Ser.*, No. 1, Vol. 1, D2 (1975)
77 HIGHLEY, J., *Inst. Fuel Symp. Ser.*, No. 1, Vol. 2, 37 (1975)
78 BORGWARDT, R.H. and HARVEY, R.D., *Environ. Sci. Tech.*, **6**, 350 (1972)
79 FIELDES, R.B., BURDETT, N.A. and DAVIDSON, J.F., *Trans. I. Chem. Engrs.*, **57**, 276 (1979)
80 O'NEILL, E.P. and KEAIRNS, D.L., *A.I.Ch.E. Symp. Ser.*, **73**(161), 100 (1977)
81 KEAIRNS, D.L., NEWBY, R.A., LANCASTER, B.W. and ARCHER, D.H., *Inst. Fuel Symp. Ser.*, No. 1, Vol. 1, D1 (1975)
82 MOSS, G., *Proc. 1st Int. Conf. Fluidized Bed Combustion*, US-EPA (1968)
83 POTTER, A.E., *Ceram. Bull.*, **48**, 855 (1969)
84 LEE, D.C., HODGES, J.L. and GEORGAKIS, C., *Chem. Engng. Sci.*, **35**, 302 (1980)
85 BORGWARDT, R.H., *Environ. Sci. Tech.*, **4**, 59 (1970)
86 YANG, R.T., GUNNINGHAM, P.T., WILSON, W.I. and JOHNSON, S.A., *Adv. in Chem. Ser.*, **139**, 149 (1975)
87 SZEKELY, J. and EVANS, J.W., *Chem. Engng. Sci.*, **25**, 1091 (1970)
88 SZEKELY, J. and EVANS, J.W., *Chem. Engng. Sci.*, **26**, 1901 (1971)
89 PIGFORD, R.L. and SLIGER, G., *Ind. Eng. Chem. Proc. Des. Dev.*, **12**, 85 (1973)
90 HARTMAN, M. and COUGHLIN, R.W., *A.I.Ch.E. Jl*, **22**, 490 (1976)
91 RAMARCHANDRAN, P.A. and SMITH, J.M., *A.I.Ch.E. Jl*, **23**, 353 (1977)
92 CHROWSTOWSKI, J.W. and GEORGAKIS, C., *ACS Symp. Ser.*, **65**, 225 (1978)
93 GEORGAKIS, C., CHANG, C.W. and SZEKELY, J., *Chem. Engng. Sci.*, **34**, 1072 (1979)
94 FIELDES, R.B. and DAVIDSON, J.F., *A.I.Ch.E. National Meeting* (Nov. 1978)
95 ZHENG, J., YATES, J.G. and ROWE, P.N., *Chem. Engng. Sci.*, **37**, 167 (1982)
96 ELGIN, D.C. and PERKS, H.R., *Proc. 5th Synth. Pipeline Gas Symp.*, AGA Cat. No. L51173, p. 145 (1974)
97 COOKE, M.J. and ROBSON, B., *Chem. Engineer*, 729 (1978)
98 KUNII, D. and LEVENSPIEL, O., *Fluidization Engineering*, Wiley, New York (1969)
99 LOM, W.L. and WILLIAMS, A.F., *Substitute Natural Gas*, Applied Science Publishers, London (1976)

CHAPTER 5
Miscellaneous processes

5.1 Apologia

The processes considered in Chapter 3 and 4 are ones that have attracted considerable interest from the scientific community and as a result are reasonably well documented in the literature. There are however many fluidized-bed processes that, largely for reasons of industrial secrecy, have received scant attention outside company reports or patents. Since in general the former are inaccessible and the latter are misleading, the processes they refer to are difficult for a reviewer to analyse in any detail. None the less, it would not be appropriate to produce a survey of fluidized-bed chemical processes without at least a cursory mention of some of the more prominent applications of the technique and this has been attempted here. It must be emphasized however that the treatment given is necessarily somewhat superficial.

5.2 Phthalic anhydride (1,3-isobenzofurandione)

Of the three benzene dicarboxylic acids only one, phthalic acid, forms an anhydride:

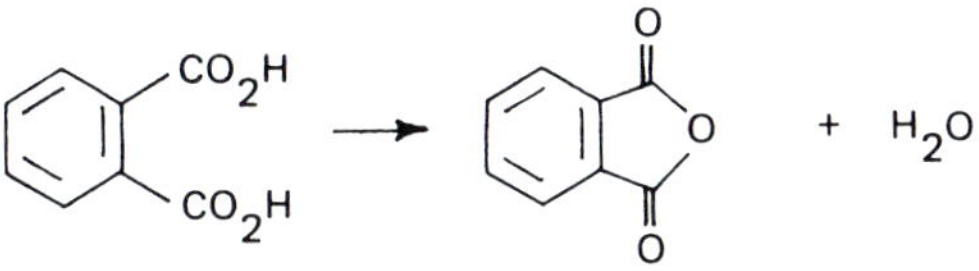

Phthalic acid Phthalic anhydride

Phthalic anhydride is a heavy organic chemical of widespread importance and has been used for a variety of purposes on an industrial scale for many years. Its earliest application was in the dyestuffs industry in the manufacture of indigo, eosin, the rhodamines and, later, the phthalocyanins. A second and higher tonnage market for it was created in the 1920s with the development of the alkyd resins used in paints and surface coatings, and although these still account for about one third of the world production, its main area of application today, accounting for some 45 per cent of the total, is in the manufacture of the phthalate esters used as plasticizers.

Flexible PVC for instance can contain as much as 25 per cent w/w of a plasticizer such as dioctyl phthalate. Another outlet for phthalic anhydride is in the production of polyester resins suitable for use with glass fibre and other reinforcing materials.

The first patented process for the manufacture of phthalic anhydride was commercialized by BASF in 1896 and involved the oxidation of naphthalene with sulphuric acid catalysed by mercury. The catalytic action of vanadium pentoxide (V_2O_5) on the air oxidation of naphthalene at elevated temperatures was discovered almost simultaneously in 1916 by Wohl in Germany and by Gibbs and Conover in the USA, and it formed the basis of all phthalic anhydride production processes in operation between 1920 and the late 1940s.

These vapour-phase processes generally employed fixed-bed reactors which were notoriously difficult to operate at full capacity and were susceptible to fire, explosion, catalyst deactivation and severe corrosion[1]. It was partly in an effort to circumvent these problems that fluidized-bed systems based on naphthalene oxidation were introduced in the mid-1940s, first by Sherwin-Williams in the USA and then by United Coke and Chemicals in the UK[2]. Unfortunately, many of the early fluidized-bed plants performed no better than those using fixed beds and it was largely as a result of plant failures coupled with a rapidly growing demand for plasticizers that the phthalic anhydride market particularly during the 1950s and 1960s was subjected to frequent and erratic fluctuations in supply.

These difficulties were aggravated by the steady decrease in the availability of coal-tar naphthalene, the long term supply of which, in Europe particularly owing to the rundown of the coal-gas industry there, looked at that time to be very uncertain. Naphthalene from petroleum sources became available in the USA in 1961 and by virtue of its lower sulphur content sold at a premium over the coal-tar product. Nevertheless, this so-called petro-naphthalene was soon faced with competition from an entirely new raw material, o-xylene, produced on an increasing scale in aromatics plants by the catalytic reforming of petroleum naphtha. The use of o-xylene for phthalic anhydride manufacture was, however, complicated by the greater amount of heat released in its oxidation and by the unsuitability of the known catalysts, so that it could not be substituted directly for naphthalene in existing plants. Furthermore its supply was linked to the demand for p-xylene to which it can be isomerized and since the demand for polyesters derived from p-xylene (via terephthalic acid) was growing rapidly, o-xylene was itself frequently in short supply[3].

The complexity of the feedstock supply situation strengthened the need for flexible processes capable of operating with more than one raw material and it was here that fixed-bed technology moved ahead. It was found that the catalyst most widely used in fluidized-bed reactors was quite unsuitable for the partial oxidation of o-xylene, whereas the fixed-bed catalysts could be adapted successfully to this feestock[1]. Furthermore, considerable improvements were made to the fixed-bed processes in the 1960s which, coupled with their increased size, and hence lower unit costs, led to the growth of their dominance[4]. The last fluidized-bed phthalic anhydride plant in the UK was shut down in 1972.

All industrial catalysts used for the air oxidation of naphthalene contain vanadium pentoxide on a moderately high surface area support (about $40\,m^2/g$) such as silica gel, alumina, magnesia or titania. The parallel advancement of fixed-bed processes in Germany and in the USA during the period 1920 to 1950 resulted in the development of two distinct types of catalyst: the 'German' type in which the activity of the V_2O_5 was modified by the incorporation of a considerable amount of potassium oxide, and the 'American' type of unmodified and highly active V_2O_5. The lower activity of the German catalyst necessitated the use in a reactor of a long contact time, and hence a low space velocity, but this was compensated for by its long life and high selectivity at low temperatures (400°C). The American catalyst operated best at higher temperatures (500–600°C) and with high space velocities; however, despite the good conversions it gave, its yield of phthalic anhydride was usually some 10 to 15 per cent lower than that obtained with the German type[5]. At a time when naphthalene was available cheaply the lower unit costs attainable with the American type gave an economic advantage to its use, but as raw material costs increased the higher yield of the German type brought this catalyst more into favour.

To be suitable for use in a fluidized-bed reactor catalysts must be resistant to attrition. The earlier catalysts for naphthalene oxidation were deficient in this respect and the fine particles they generated were the cause of frequent problems either through their contamination of the condensed product (phthalic anhydride is a solid at ambient conditions), or, by depositing on filters in the reactor exit lines, catalysing the further oxidation of the anhydride. Catalysts prepared by wet impregnation or gel precipitation methods are particularly prone to attrition but a method developed in the 1950s gives much better materials. This involves fusing the V_2O_5 with potassium pyrosulphate ($K_2S_2O_7$) thereby forming a glass which is then powdered and added to a fluidized bed of the support material maintained at 350–500°C. Under these conditions the glass melts and sticks to the support surface forming a coating which remains intact during its subsequent use in the reactor.

In a typical example of a fluidized-bed phthalic anhydride process, the bed of catalyst is fluidized with air and liquid naphthalene is pumped from storage to enter the bed through spray nozzles located just above the air distributor. The reactor temperature, 340–380°C, is controlled by using the exothermic heat of reaction to raise high-pressure steam through tubular heat exchangers immersed in the bed. The product passes through filters in the roof of the reactor to a separation train where the crude product is condensed out, up to 60 per cent in liquid form using conventional shell-and-tube condensers, and the remainder as a solid in switch condensers operated in condense-melt cycles. The crude product is then purified by vacuum distillation. Riley[6] has given a review of much of the early work on the use of fluidized-bed reactors in this field.

5.3 Acrylonitrile (prop-3-enenitrile)

In contrast to the relative failure of fluidized-bed technology in the manufacture of phthalic anhydride, virtually the whole world production of acrylonitrile, some four million tonnes per annum, is made in fluidized-bed reactors. Acrylonitrile first came into prominence as a heavy organic chemical in the 1930s when it was used as a copolymer with butadiene in the manufacture of synthetic rubber (nitrile-rubber or buna-N). Early manufacturing processes were based on the addition reaction between ethylene oxide and HCN (a two-stage process) and a one-stage combination of HCN with acetylene. The latter, despite its disadvantageous use of increasingly expensive raw materials, was the major synthetic route until 1960 when it was rapidly supplanted by the elegant one-stage propene ammoxidation process that has retained its dominance to the present day.

The catalysis by bismuth molybdate ($Bi_2O_3 + MoO_3$) of the reaction between propene, ammonia and oxygen:

$$CH_3CH = CH_2 + NH_3 + \tfrac{3}{2}O_2 \rightarrow CH_2 = CHCN + 3H_2O;\ \Delta H = -515\ \text{kJ/mol}$$

was patented by Idol[7] of Sohio in 1959 and the first large-scale plants were based on the application of this discovery. Subsequently catalysts based on oxides of antimony and uranium ($Sb_2O_4 + UO_3$), antimony and iron ($Sb_2O_4 + Fe_2O_3$) and antimony and tin ($Sb_2O_4 + SnO_2$) were employed giving improved yields and selectivity, but the catalyst used currently on a wide scale (catalyst C41) is reported to be a multicomponent mixture containing bismuth, molybdenum, alkali metal, alkaline earth, nickel or cobalt and phosphorus and/or arsenic and antimony[8]. The active catalytic components are normally supported on silica to give them attrition resistance in fluidized beds. The chemistry and catalytic action of these materials has been discussed in detail by Gates *et al.*[9].

Although fixed-bed reactors have been used for the ammoxidation process the most common reactors in current use are fluidized beds based on Sohio technology. The reasons for their preference are mainly:

(i) The high rate of heat evolution from the exothermic reaction (some 20 MJ/kg of acrylonitrile) requires a very efficient cooling system to maintain bed temperatures in the required range of 400–510°C. Cooling coils arranged horizontally within the fluidized bed and located at regular intervals enable this control to be achieved efficiently[10].
(ii) Both reactants and products are flammable with air within wide limits (*Table 5.1*) but a fluidized bed, because of the flame quenching action

Table 5.1 Ignition temperatures and explosion range for main reaction components of the acrylonitrile synthesis (from Sax[11])

Compound	*Ignition temperature* (K)	*Explosive range* (% by volume in air)
propene	770	2–11
ammonia	924	15–28
acrolein	551	2.8–31
acrylonitrile	754	3.1–17

of the moving particles, enables the three reactants to be admitted together into the reactor without hazard.

(iii) The life of the catalyst can be maintained without regeneration for a period of some years provided its temperature is kept below about 530°C. Again, the good temperature control possible with fluidized beds makes this feasible.

The general arrangement of the reactor is shown in outline in *Figure 5.1.* Operating temperatures and pressures are in the ranges 400–510°C and 1–3 bar respectively. Callahan *et al.*[13] gave a value of the apparent contact

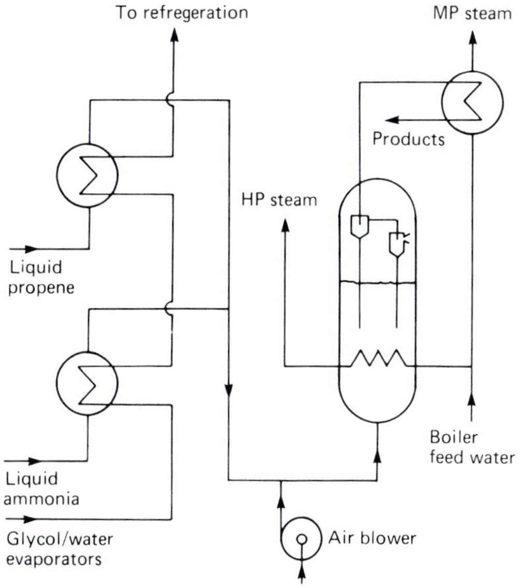

Figure 5.1 Reactor layout for acrylonitrile synthesis by ammoxidation of propene (From Pujado et al.[12])

time, (volume of catalyst in the bed divided by the volumetric flowrate of reactants), of 1–15 s, and feed ratios for the C41 catalyst of:

oxygen to propene	1:1 to 3:1
ammonia to propene	0.5:1 to 5:1
steam to propene	1:1 to 4:1

In addition to acrylonitrile a number of by-products are produced, among which are HCN, acetonitrile, oxides of carbon and acrolein. The newer

C41 catalyst, which is a microspheroidal powder with an average particle size of about 60 μm, gives a lower yield of by-products as well as having a higher reactivity than the older types.

5.4 Vinyl chloride (chloroethene)

Chlorinated hydrocarbons find widespread use in industry as solvents, intermediates in the production of other halogenated materials such as chlorofluorocarbons, and principally in the form of vinyl chloride monomer as the starting material for polyvinyl chloride. Vinyl chloride is said to be the fourth largest tonnage chemical produced in the wrold and outside petroleum refining the largest tonnage organic product of a catalytic process[8]. Before the advent of cheap petrochemicals in the 1960s vinyl chloride was made from acetylene (ethyne) and HCl in a one-stage process:

$$CH = CH + HCl \rightarrow CH_2 = CHCl$$

but as acetylene became more expensive relative to ethylene (ethene), processes based on the latter material were developed. A number of those in current use employ fluidized-bed reactors in one of the stages of a generally three-stage route. The first stage is one normally carried out in the liquid phase with 1,2-dichloroethane (ethylene dichloride or EDC) as solvent and ferric chloride as catalyst. Ethylene and chlorine gas are admitted to the system at 40–100°C and 1–10 bar to give high yields of EDC:

$$CH_2 = CH_2 + Cl_2 \rightarrow CH_2ClCH_2Cl$$

In the second stage the EDC is cracked by heating in a tubular reactor at 450–600°C and 10–35 bar:

$$CH_2ClCH_2Cl \rightarrow CH_2 = CHCl + HCl$$

From an economic point of view it is essential to utilize the HC1 by-product from the second stage and it is in the realization of this objective that a fluidized-bed reactor is usefully employed. Ethylene, HC1 and air or oxygen are passed through a bed of alumina-supported cupric chloride catalyst at 200–300°C and 2–10 bar. A highly exothermic reaction occurs in which further EDC is produced:

$$CH_2 = CH_2 + 2HCl + \tfrac{1}{2}O_2 \rightarrow CH_2ClCH_2Cl + H_2O; \Delta H = -2O1 \text{ kJ/mol}$$

In this oxychlorination reaction the bed temperature must be carefully controlled in order to prevent (i) combustion of the ethylene, and (ii) overchlorination of the EDC product. The use of a fluidized bed enables this control to be achieved giving overall yields and selectivities based on ethylene in excess of 90 per cent[8].

5.5 Titanium dioxide

Titanium dioxide is the most important of all white pigments and is used on a large scale in the paint, plastics and paper industries. It exists in two crystalline forms, rutile and anatase and although each can be made industrially depending on the processing conditions used, the majority of pigments currently produced are in the rutile form. The two major ores from which the material is obtained are ilmenite containing some 40–70 per cent TiO_2, and rutile which is about 95 per cent TiO_2; oxides of iron make up the majority of the balance. Known reserves of rutile are quite small and ilmenite, either natural or beneficiated, is the main raw material of the industry, although considerable quantities of slags, derived from the electrosmelting of ilmenite in pig-iron production, are also used. The two processes most widely employed for the production of pigment-grade titanium dioxide are (i) the wet sulphate process, in which ground ilmenite or slag is digested with concentrated sulphuric acid to give a solution of sulphates of titanium and iron from which the $FeSO_4$ is crystallized out leaving the $TiSO_4$ to be calcined to TiO_2, and (ii) the more recently developed dry chloride process in which fluidized-bed reactors are used.

In the chloride process, in order to minimize the usage of expensive chlorine, the TiO_2 content of the ore is frequently increased to over 90 per cent by beneficiation using a leaching technique. The process then involves reacting the beneficiated material with chlorine in the presence of a reducing agent such as coke to give titanium tetrachloride:

$$2TiO_2 + 4Cl_2 + 3C \rightarrow 2TiCl_4 + CO_2 + 2CO$$

The reaction is carried out at a temperature of about 950°C in a fluidized bed of refractory-lined carbon steel. The reason for choosing this type of system rather than a fixed-bed arrangement, which was used previously, is largely due to the good solids mixing characteristics of the fluid bed which enable ore and coke to be brought into intimate contact under reaction conditions. In fixed-bed reactors it is necessary first to bind the two solids together into briquettes before the chlorination[14].

The product stream from the fluidized-bed reactor is cooled to remove some of the impurities as dry solids leaving the remaining vapours to be condensed. The $TiCl_4$ is separated by fractional distillation before being burned with oxygen or oxygen-enriched air at 1400–1500°C to give the dioxide:

$$TiCl_4 + O_2 \rightarrow TiO_2 + Cl_2$$

A fuller description of both pre- and after-treatment processes was given by Darby and Leighton[15] who also quoted world production figures for 1980 of TiO_2 by the sulphate and chloride routes, of 1.81 million tons and 0.935 million tons respectively.

5.6 Uranium processing

The nuclear power industry requires large quantities of uranium-based fuel materials and fluidized beds are used in several stages of their production. A summary of the main types of reactor in current use is given in *Table 5.2*,

Table 5.2 Characteristics of major reactor types (from Findlay *et al.*[16])

Reactor	*Magnox*	*Advanced gas cooled (AGR)*	*Light water (pressurised and boiling) (PWR & BWR)*	*Canadian deuterium moderated (CANDU)*	*Water/ graphite*	*High temperature (HTR)*
Country of origin	Britain France	Britain	USA	Canada	USSR	Several
Fuel material	Metal	Oxide	Oxide	Oxide	Oxide	Carbide (or oxide)
Fuel form	Rod	Pellet	Pellet	Pellet	Pellet	Spherical particle
Fuel enrichment	Natural U 0.7% U235	2% U235	3–4% U235	Natural U	2% U235	93% U235
Cladding	Magnox	Stainless steel	Zircaloy	Zircaloy	Zirconium Niobium alloy	Graphite/ silicon carbide
Moderator	Graphite	Graphite	H_2O	D_2O	Graphite	Graphite
Coolant	CO_2	CO_2	H_2O	D_2O	H_2O	Helium

from which it is clear that the two most commonly employed fuels are uranium metal and uranium oxide. The Magnox reactors developed in the UK in the late 1940s are fuelled by uranium metal produced by a series of chemical reactions starting from a uranium ore concentrate[17]. The latter is leached with nitric acid to give a solution of uranyl nitrate which is then purified by filtration and solvent extraction. The uranyl nitrate is concentrated to the hexahydrate by evaporation to give a thin liquid with a freezing point of 78°C, which may be converted to uranium trioxide by thermal decomposition at 300–400°C:

$$UO_2(NO_3)_2 \rightarrow UO_3 + NO_2 + NO + O_2; \Delta H = 332 \text{ kJ/mol}$$

This endothermic reaction is carried out by spraying either uranyl nitrate solution[18] or the hexahydrate in molten form[17] into a bed of UO_3 particles fluidized with air. The particles grow in size as the nitrate decomposes and deposits fresh UO_3 on to them. (Hawthorn *et al.*[17] described the development of a pilot-scale denitration plant 9 in. in diameter and 3 ft deep with a power requirement of 20 kW which was provided by electrical resistance heating. A similar scheme operated in the USA was discussed by Bjorklund and Offutt[18]). Enlarged particles of UO_3 are then withdrawn and transferred either batchwise or continuously to a second reactor where they are fluidized with a mixture of hydrogen and nitrogen at temperatures in the region of 650°C. Under these conditions the trioxide is reduced to UO_2:

$$UO_3 + H_2 \rightarrow UO_2 + H_2O; \Delta H = -100 \text{ kJ/mol}$$

A third reaction stage in which the UO_2 is hydrofluorinated to UF_4 with HF gas is also carried out in a fluidized system although again, because of the endothermic nature of the reaction, external heating is required, to give a temperature of about 550°C:

$$UO_2 + 4HF \rightarrow UF_4 + 2H_2O; \Delta H = 180 \text{ kJ/mol}$$

The design of hydrofluorination plant described by Hawthorn *et al.*[17] required two vessels each with two beds in series for continuous operation. The lower rate of the hydrofluorination reaction compared with that of the preceeding UO_3 reduction also necessitated a longer solids hold-up and a consequent use of deeper beds. Fine particles of UO_2 have a tendency to sinter and stick to the walls of the reactor and systems have been described in which this is prevented by agitating the bed with a slow-speed stirring device[19]. Uranium metal for Magnox fuel elements is produced in a variant of the thermite process by reacting UF_4 with magnesium in graphite-lined steel vessels:

$$UF_4 + 2Mg \rightarrow U + 2MgF_2$$

Parallel American developments were described by Brater *et al.*[19] and by Philoon *et al.*[20] who gave the following reasons for choosing fluidized beds for the foregoing conversions:

(i) the ease of maintaining uniform temperatures in the reactors,
(ii) the good heat transfer properties for both endothermic and exothermic reactions,
(iii) the use of a closed system which minimized the risk of exposure of personnel to radioactive dust.

The early stages of the process for making uranium oxide fuel are similar to those described above for the metal. Here however, because of the need to enrich the fuel with ^{235}U it is necessary first to proceed to the gaseous uranium hexafluoride, UF_6 via fluorination of UF_4:

$$UF_4 + F_2 \rightarrow UF_6$$

The hexafluoride, after treatment in an enrichment plant, is converted to UO_2 powder by one of a number of methods. A widely used dry method involves reacting UF_6 with superheated steam to give the solid uranyl fluoride (UO_2F_2) which is then reduced to UO_2 by reaction with hydrogen in either a fluidized bed or a rotary kiln. The final stage of preparation of the fuel involves the production of pellets of UO_2 by cold pressing followed by sintering in hydrogen at 1600–1700°C[16].

Another type of nuclear fuel produced for use in helium-cooled high temperature reactors (HTR), uses uranium carbide, UC_2, or uranium-thorium carbide, $UThC_2$, particles coated with a form of graphite. The coating process is carried out in a fluidized bed of fuel particles supplied with a hydrocarbon such as methane or acetylene (ethyne) in an inert carrier gas and maintained at a temperature in the range 1200–2000°C. Under these conditions the hydrocarbons 'crack' and deposit a form of graphite on the particles. The structure of this pyrolytic graphite depends on the nature of the hydrocarbon and of the reaction temperature[21,22] but in all cases it provides a strong coating to the particles, enabling both fuel and fission products to be retained in a coherent form throughout the reactor cycle. A coating of silicon carbide can be produced in the same way by the use of methyl trichlorosilane, CH_3SiCl_3[16].

5.7 Sulphide ore roasting

Fluidized beds have been used for the roasting of sulphide ores such as pyrite, FeS_2, since the early 1950s when the technique was developed independently by Dorr–Oliver in the USA, Badische Analine (BASF) in Germany and Sumitomo in Japan[23]. The exothermic oxidation reaction of the sulphide is carried out by fluidizing ore particles under 10 mm in size with air in large diameter beds at temperatures in the range 650–1100°C:

$$4FeS_2 + 11O_2 \rightarrow 2Fe_2O_3 + 8SO_2$$

The SO_2 product is used for sulphuric acid manufacture. Kunii and Levenspiel[23] gave the following figures for a Dorr–Oliver roaster of internal diameter 5.5 m and height 7.6 m:

pyrite feed rate	170–220 ton/day
bed temperature	650–700°C
bed height	1.2-1.5 m
SO_2 in exit gas	12%

Start up of these units is carried out normally by gas firing and temperature control during operation is either by top spraying with water or heat exchange via immersed cooling coils. The BASF Turbulent Layer process, engineered by Lurgi (*Figure 5.2*) uses a shallower bed, 0.6 m deep, run at higher temperatures with gas velocities in the range 1–3 m/s[24].

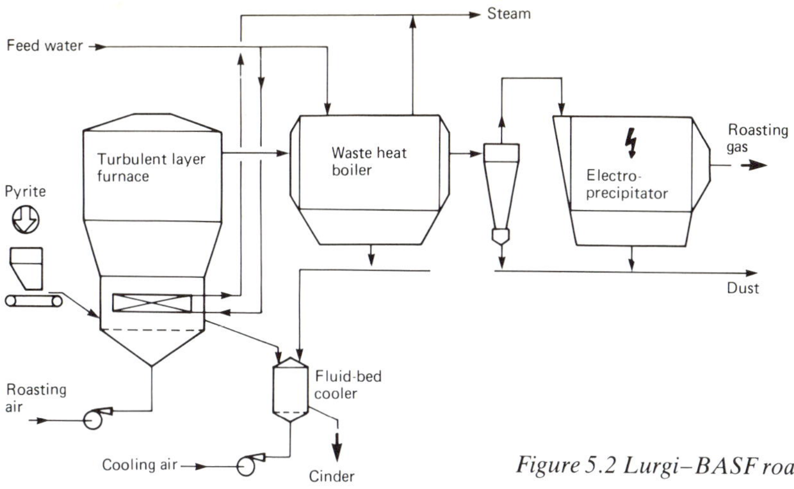

Figure 5.2 Lurgi–BASF roaster

Large numbers of these systems have operated throughout the world and many of their reactor units are of considerable size although their economic viability as producers of SO_2 has been increasingly threatened by the growing availability of elemental sulphur from petroleum refining and petrochemicals manufacture. The cost of a roasting plant can be more than twice that of a sulphur-burning plant and, coupled with the generally higher maintainance costs of roasters, it often tips the balance of favour against them even when an indigineous supply of pyrite is available[25]. The largest fluidized-bed roaster in the world is said to be that built by Lurgi in Tasmania; it has a grate diameter of 12.5 m and a height of over 16 m[24].

5.8 Direct reduction of iron ores

The two main methods used in the production of iron for steelmaking are the so-called cold-metal and hot-metal routes. In the former a material such as scrap iron is melted in an electric furnace and converted *in situ* under controlled conditions. The hot-metal route employs a blast furnace which is fed with a mixture of prepared iron ore, limestone and coke. The coke serves the dual purpose of providing the reducing gases carbon monoxide and hydrogen that remove oxygen from the ore, and also of burning within the furnace so maintaining its temperature at around 900°C and allowing the iron metal to be removed periodically in molten form. In recent years an alternative 'hot' method has been developed which does not use a blast furnace. This is the 'direct reduction' process in which the ore is reacted in the solid form at alower temperatures with a reducing gas mixture produced by the steam reforming of a material such as natural gas. The process gives a solid product suitable as an alternative to scrap for treatment in an electric furnace.

The central reactor of the direct reduction route where the ore conversion is carried out may take one of a variety of forms: a vertical moving bed in the Midrex and ARMCO processes, and a fluidized bed in the H-Iron and FIOR processes [26]. The H-Iron process, developed in the USA by Hydrocarbon Research and Bethlehem Steel, uses multistage fluidized beds at 490–550°C mounted one above the other, with iron oxide powder being introduced at the top and moving via downcomers to the lower beds against a countercurrent flow of reducing gas. The FIOR process developed by Esso (now Exxon) uses multiple fluidized beds.

Although these direct reduction methods do not yet pose a serious threat to the blast furnace as the main reactor for hot-metal processes, their use in the future could well become more widespread. Not only are they capable of being operated economically on a smaller scale than that of the blast furnace, an important factor in developing countries without the capital resources essential for large integrated steel works, but also their independence from the dwindling supplies of good quality coking coals is a considerable advantage. Moreover the attractiveness of direct reduction methods for the developed nations lies in the possibility of linking them to electric furnaces supplied with cheap nuclear-generated electricity, and ultimately to nuclear reactors that provide a source of direct heat[27]. The many advantages of fluidized beds for solids handling under reaction conditions makes it highly probable that they will have a major role to play in these future developments.

References

1 SPITZ, P.H., *Hydrocarbon Proc.*, **47**(11), 162 (1968)
2 GRAHAM, J.J. and WAY, P.F., *Chem. Engng. Prog.*, **58**(1), 96 (1962)
3 DUCKWORTH, R.A., *Chem. Proc. Eng.*, No. 1, 69 (1969)
4 OCKERBLOOM, N.E., *Hydrocarbon Proc.,* **50**(9), 162 (1971)
5 LANDAU, R. and HARPER, H., *Chem. Ind.*, 1143 (1961)
6 RILEY, H.L., *Trans. I. Chem. Engrs.*, **37**, 22 (1959)
7 IDOL, J.D., US Patent 2904580 (1959)
8 PEARCE, R. and PATTERSON, W.R., *Catalysis and Chemical Processes*, Leonard Hill, Glasgow (1981)
9 GATES, B.C., KATZER, J.R. and SCHUIT, G.C.A., *Chemistry of Catalytic Processes*, McGraw-Hill, New York (1979)
10 VAN DER BAAN, H.S., in *Chemistry and Chemical Engineering of Catalytic Processes*, Eds. Prins, R. and Schuit, G.C.A., *Nato Advanced Study Institute Series E:* Applied Sciences No. 39 (1980)
11 SAX, N.I., *Dangerous Properties of Industrial Materials*, 4th Edn, Van Nostrand Reinhold, New York (1975)
12 PUJADO, P.R., VORA, B.V. and KREUDING, A.P., *Hydrocarbon Proc.*, **56**(5), 169 (1977)
13 CALLAHAN, J.L., GRASSELLI, R.K., MILBERGER, E.C. and STRECKER, H.A., *Ind. Eng. Chem. Proc. Des. Dev.*, **6**, 134 (1970)
14 GELDART, D., *Chcm. Ind.*, 41 (1968)
15 DARBY, R.S. and LEIGHTON, J., in *The Modern Inorganic Chemicals Industry*, Ed. Thompson, R., Chemical Society, London (1977)
16 FINDLAY, J.R., GLOVER, K.M., JENKINS, I.L., LARGE, N.R., MARPLES, J.A.C., POTTER, P.E. and SUTCLIFFE, P.W., in *The Modern Inorganic Chemicals Industry*, Ed. Thompson, R., Chemical Society, London (1977)

17 HAWTHORN, E., SHORTIS, L.P. and LLOYD, J.E., *Trans. I. Chem. Engrs.*, **38**, 197 (1960)
18 BJORKLUND, W.J. and OFFUTT, G.F., *A.I.Ch.E. Symp. Ser.*, **69**(128), 123 (1973)
19 BRATER, D.C., CASTRO, D.L., DEAN, L.A., PASHLEY, J.H. and SMILEY, S.H., *Chem. Eng. Prog. Symp. Ser.*, **62**(65), 1 (1966)
20 PHILOON, W.C., SANDERS, E.F. and TRASK, W.T., *Chem. Engng. Prog.*, **56**(4), 106 (1960)
21 CULLIS, C.F. and YATES, J.G., *Acta Cryst.*, **17**, 1433 (1964)
22 CULLIS, C.F. and NORRIS, A.C., *Carbon*, **10**, 525 (1972)
23 KUNII, D. and LEVENSPIEL, O., *Fluidization Engineering*, Wiley, New York (1969)
24 REH, L., SCHMIDT, H.W., DARADIMOS, G. and PETERSEN, V., *Inst. Energy Symp. Ser.*, No. 4, Paper 6-2 (1980)
25 PHILLIPS, A., in *The Modern Inorganic Chemicals Industry*, Ed. Thompson, R., Chemical Society, London (1977)
26 KUNII, D., *Chem. Engng. Sci.*, **35**, 1887 (1980)
27 FINNISTON, M., *Chem. Ind.*, 501 (1976)

Author index

Subject index